The Great Realignment: Power, Money, Greed & Bitcoin

D.L. WHITE

Copyright © 2024 D.L. White

All rights reserved.

ISBN: 979-8-218-51486-0

DEDICATION

To Professor Tim Newburn
Thank you.

CONTENTS

	Preface	i
	Introduction	1
1	PART ONE: Power	5
2	Strength	9
3	Control	22
4	Authority	35
5	Force	40
6	PART TWO: Money	51
7	Commodity Money	60
8	Fiat Money	69
9	Other Money	82
10	PART THREE: Greed	98
11	Choices	110
12	Chaos and Luck	122
13	Waste	131
14	PART FOUR: Bitcoin	147
15	Alts and CBDCs	155
16	The Great Realignment	171
17	AI, War and the Future	187
	References	195

PREFACE

This book was written "open-source." Meaning I published everything online as I wrote. I cannot be certain this is a unique, or indeed, even a sound approach to writing a book. Yet, it is the approach I decided upon. For those early readers, this was unlike most books, where the effort to compose it is long since past by the time the reader puts their eyes to the page. The early readers were along for the ride, so to speak. Just prior to writing this I realized I have written 350,000 to 400,000 words on Medium over the last two and a half years. In terms of most mass-market books, that word count falls somewhere between three to five books worth of material. To be fair, most of it would not be worthy of a book. My approach on Medium was largely devoted to an educational campaign around basic macro-economic concepts and the safe navigation of digital asset markets. Of late that focus has transformed to a singular one: The promotion of Bitcoin as an inalienable property right and reserve asset.

This shift in endeavor did not come overnight, nor was it one that was undertaken lightly. Prior to writing on Medium, I published a number of peer-reviewed academic papers on a diversity of topics. While I am a firm believer that an advanced education and defense of ideas is a noble pursuit, I recognize now that such endeavors come with hidden costs. Most notably, the rigor I applied to evaluating so-called "crypto assets" blinded me to a number of unique and distinguishable properties of Bitcoin. The academic mind is a suspicious one and rightfully so. Extraordinary claims must be backed by extraordinary evidence. Absent such backing, caution is always warranted before one draws strong conclusions.

Unfortunately, this led me to a number of unseen and unacknowledged biases. The conservatism of academia provided me with an intellectual cover that I ego-centrically pursued to maintain an appearance of objectivity. Ego-centric or not, objectivity is appropriate when discussing topics such as those found in scholarship on financial regulation and criminology—my particular areas of expertise. Within those fields, there is a vast history of critical thinkers and astute, learned scholars working through the often intractable problems of human interaction. Novelty of approach and analysis is uncommon, as the depth of existing scholarship rarely has holes large enough to find a new idea worth exploring.

Not because seemingly new ideas do not materialize. More so because what can often be mistaken for a new idea is rarely one that has not been thoroughly discussed, debated or definitively answered by someone, or indeed, many someones along the way. This being a long-winded way of saying, there are very few unturned stones in legal and criminological circles. This is why academic papers are so often focused on minutia and feature much tinkering around the edges of core ideas. Thus the introduction of something that is truly new can be disruptive. The impulse is to reject novelty out of hand. Not due to a lack of curiosity, but rather because curiosity in academia is so rarely rewarded.

Again, conservatism of approach is highly prized and well regarded in that field. To take a stand on a new concept is to challenge much orthodoxy and it is an uncomfortable place to remain steadfast. And yet, my curiosity continued to pursue the little threads and tendrils of Bitcoin knowledge and ideas that are often presented as irrefutable or inevitable by the devout. Unfortunately the certainty of the Bitcoin devotees was anathema to my predisposition, right or wrong. All of which conspired to make the revelation of Bitcoin's deeper qualities very opaque to me. I consider myself fortunate that I set aside my suspicion long enough to pursue greater knowledge of Bitcoin with an open mind.

Upon doing so, it became evident that I had been wrong about Bitcoin all along. That said, I think my true blessings are the possession of a sufficient quantity of humility to recognize I was wrong and the ability to correct course. This book is my attempt to contribute to, explore and expand Bitcoin knowledge. My intention is to bridge an undefinable gap between "pure" academic writing and mass market narration. My hope is to achieve some form of middle ground. The ultimate goal is to provide an alternate reference point for discussion and debate. If this proves limiting to my audience, so be it. My incentive is pure. I am solely focused on enhancing and furthering constructive dialogue around Bitcoin and what I believe is its primary utility as inalienable individual property.

I am certain many will disagree with my analysis and it is welcome criticism. I am not, and will do my best not to become, dogmatic in approach. In terms of sourcing and citation, this was originally intended for a digital medium. My idea was to link to sources. As I have decided to also publish a physical print book, bibliographical references are now included as well. There are two-fold issues here nevertheless. The trouble with links is they can move or expire. The trouble with bibliographies is they are rarely pursued, as I learned all too well when I was called upon to be a peer-reviewer myself. That being said, this work is very modestly sourced. It is not intended to be a deep academic treatise and much of it is theoretical anyway. I suppose only time will tell how well it all holds up.

Shifting gears a bit and for insight into the writing of this book, I published online weekly. It was an odd process for me as a writer. This method of work certainly developed and morphed the book from where I started. Long form writing requires a lot of mental shifting. And shifting from focused to diffuse thinking has uncertain timing. For me, this is not a predictable practice. It is one, however, that is critical to navigating a long writing assignment successfully. All this to say, the writing I did over these last few months was not linear, nor was it complete. It was really quite fractured because of my intention to post updates to the book weekly, regardless of the state of things. This was all in pursuit of a secondary hope I held for this endeavor. That hope was it might benefit burgeoning writers with their journey. I thought that providing an example of the writing process unfolding week-by-week might prove helpful. Of this I saw little impact. It was a fun experiment to try, despite the lack of results. That said, I completed this book in its entirety on Medium. From here, it shall be published (hopefully) on Highlighter.com via the NOSTR protocol. It will also be published on Amazon for those who might like to buy a physical copy. I will leave the choice to the consumer. Selling the book is only for the convenience of those who do not like reading from a website, but are still interested in what I have to say on the matter at hand.

I am a firm believer that Bitcoin will be one of the most profound, fundamental changes in the way human beings interact and transact. That being said, I also find the idea that Bitcoin's greatest purpose is a medium of exchange for any and all transactions a bit absurd. This puts me in some conflict with many of the original developers that helped create the Bitcoin we know today. Regardless, the amount of infighting and hostility among the various Bitcoin republics is disturbing to me. I think most of it childish and counter-productive to the potential embodied in the small-block version of Bitcoin. That is to say, the version that is currently preeminent among the ideologies, at least in terms of adoption. A perhaps vain hope for this book is to eventually bridge the gap and find unity among the various camps. As I alluded to above, the implications of Bitcoin as a treasury reserve asset and individually sovereign property seem to me far more important and impactful than a censorship-resistant, black market, micro-transactional network could ever hope to be.

In pursuit of this argument, this book advances a number of historical theories, namely around the concepts and origins of power, money and greed in our modern economy. I do this to provide an alternate view of the world molded by my own academic experience and work. This is not intended to upset any apple carts. The purpose is rather to provide more context for the broader economic discussion and where this Bitcoin thing might end up. I think the Bitcoin story, much like the human story, is truly about to blossom. My hope is to present a coherent and hopeful narrative along which those stories might evolve.

INTRODUCTION

I hope you will indulge me for a moment before we begin the broader discussion. The content of this book will likely be discomforting to many. I am sure many sacred cows are sacrificed within the following pages. Likewise, those pages will undoubtedly challenge a number of orthodoxies. This is not simply for sport. The intention throughout is to highlight potential sources of error and disgorge unfounded assumptions of their sediment. The bulk of my life is governed by a disposition and orientation that is eloquently summed by the late theoretical physicist Richard Feynman. He says:

> *I think it's much more interesting to live not knowing than to have answers which might be wrong. I have approximate answers and possible beliefs and different degrees of uncertainty about different things, but I am not absolutely sure of anything and there are many things I don't know anything about.*

Thus this book. The trouble I find with academia in general—whether it be law, history, politics, economics, or finance—is the observable reality that there is a market for the product of those disciplines. And those products are most certainly governed by market forces. Tenure track careers hinge on consistent academic publication. Lucrative financial rewards and professional esteem await the successful cross-overs to mass market literature and the speaking tour circuit. Indeed, many references in this book belong to those academics that have made the successful transition.

The trouble is, mass media and mass markets are not amenable to nuance. And once embroiled in the prestige of popularity, it is often quite difficult for those academics to change tack mid-course, even if there are nuanced reasons to do so. Given the rarity of such successes in mass-market academic crossover, it is no wonder that points of view and narratives become entrenched. For the up and coming academic, the market demands they pay tribute and homage to those atop the publication hierarchy. Failure to do so is a fast track to irrelevance. Much the same is true in the worlds of finance, entertainment, social media and politics. Those who best express themselves while staying within the contours of what "everyone" knows tend to rise to prominence. As a social lubricant, this is all well and good. Or, as William "Matt" Briggs says, "You will never get fired for being wrong in the right direction." But this is fatal to rational inquiry.

Generally speaking, I think people gravitate towards the most reasonable sounding explanation that aligns with their subjective experience. But people today are no more sophisticated than the people of the ancient world. While we might like to separate ourselves from our seemingly unrefined predecessors, the humbling reality is that, outside of our gadgets, we are truly no different than they were. We have the same biases, the same emotional and logical errors and the same blindspots. While we may take for granted that we "know" things like the germ theory of disease, or that "balancing the humours" is a failed medical practice, what we really have are snippets of information that we are only loosely familiar with. Asserting facts based on a loose familiarity with information is really just an act of faith.

The purpose of this book is two-fold. I first want to challenge common assumptions. Ideally this will help to humanize and contextualize political-economic history. Second, I want to provide some useful theories to help people better understand Bitcoin and its role in the global political economy. Ultimately this is a book about uncertainty. And it is a book that attempts to construct a view of history from a novel perspective. The base concept underlying this book is simple: Most of what we know about the world is wrong. Which is to more accurately say, most of what we *believe* we know about the world is wrong. The number of people on earth that are true experts in the fields of history, anthropology, economics, biology, psychology, political economy, political science, international relations and dozens of other disciplines probably number in the hundreds. Most that are passed off as experts are not. These so-called "experts" are rather, and much more often, ego-centric mid-wits that further a political or ideological agenda under the guise of scientific rigor.

Worse yet, under that guise of scientific rigor, many of these mid-wits will boldly lay claim to truth. While I lay no claims to truth, I acknowledge I may well be one of those ego-centric mid-wits myself. That being said, while they parade their theories on whatever subject they study as definitive accountings of the world we live in, I am simply providing the theories that follow as a counter narrative. While media savvy academics know that certitude is a soothing tonic to lay audiences, I have little certitude on offer here. For any theory to be "true" from a scientific perspective requires that every single premise and assumption made must also be provably true, using a chain of arguments that leads to axiomatic principles. All economic theories fail this test. All political science theories fail this test. Most scientific theories will fail this test. Indeed, most of the theories in this book will fail that test. That is a good thing. That is the nature of uncertainty. That is the nature of improving knowledge.

As you have probably gleaned from the title, this book is broken down into four parts: Power, money, greed and Bitcoin. While there is much history invoked in the coming pages, this is not a history book. It is rather a book about history. Or, I should rather say, the absence of history. The trouble with history is most of it is lost. There is a span of several thousand years between the discovery and widespread adoption of agriculture and the invention of the written word. This is why the history books are full of tales of great empires going to war, but very little on how those empires came to be. Once history was able to be documented, it is much easier to follow the trail. But that which came after agriculture and before the written word is left to the realms of archaeology and anthropology. Ask any of them and they will tell you point blank there is far more they do not know than they do. The genesis of the armies and the kingdoms and the great civilizations is, quite simply put, a matter for scarcely informed speculation. It is within that great chasm of knowledge that I attempt a modest exploration here.

What I can say with some certainty is that these power structures, these monetary systems and these means of human organization are not the products of nature. They are man-made. They are constructs. They are ancient. And, as best I can tell, they all follow similar and recursive trajectories throughout the ages. I think there is very little that is new under the sun. That being said, I also think Bitcoin is a very new thing under the sun. It is a monumental achievement in the 12,000 years of post-agricultural human development. I truly believe Bitcoin changes everything. What follows is the reason why.

PART ONE: POWER

CHAPTER ONE

Discussions around power reveal many of the same flaws that we will see in our discussions about money. And much like the discussion on money that follows in Part Two, it is difficult to accurately place Bitcoin in the greater scheme of things without understanding political power. Many simply assume they know what it means to be "in power," or to "have power." Similar issues arise when discussing whether a particular country or agency is "powerful." Yet, when asked to explain or define what those terms mean, I find very few can give a concrete answer. Much like money, power is an abstract concept that attempts to describe how human beings behave and how they interact. Similar to money, how you view concepts around power is heavily dependent on your starting assumptions. For a good starting point then, we should look to the plain language descriptions. Obviously, power has a number of connotations, but the definitions[1] of power pertinent to the discussion at hand are:

1. Political or national strength: *The Second World War changed the balance of power in Europe.*
2. The possession of control or command over people: *Words have tremendous power over our minds.*
3. Political ascendancy or control in the government of a country, state, etc.: *They attained power by overthrowing the legal government.*
4. Legal ability, capacity, or authority: *The legislative powers vested in Congress.*

5. Delegated authority; authority granted to a person or persons in a particular office or capacity: *A delegate with power to mediate disputes.*
6. A document or written statement conferring legal authority.
7. A person or thing that possesses or exercises authority or influence.
8. A state or nation having international authority or influence: *The great powers held an international conference.*
9. A military or naval force: *The Spanish Armada was a mighty power.*

Are you starting to see the issue? When we speak of *power* it is quite easy to assume we are talking about something specific. Once broken down, however, power can be a very nebulous word. As the old saying goes, "The biggest problem with communication is the illusion it has been accomplished." When speaking of power, there are clearly many barriers to effective communication. If we take a closer look at the definitions above, we can see some common themes. Power can and does—often interchangeably—represent:

1. Strength
2. Control
3. Authority; and
4. Force.

Yet, if we boil this down even further, it does little to help us. Speaking about *strength*, for instance, what does strength really mean? Take the subtext from the first definition of political or national strength and fill it out. "*The Second World War changed the balance of* STRENGTH *in Europe?*" Strength of what though? Moral strength? Physical strength? Defensive strength? Military strength? Diplomatic strength? Moreover, if you look at the definition of strength[2] you will find the word *power* used throughout. Let us take it a step further and think about *control* for a moment. The subtext of the next couple of definitions say, "*Words have tremendous power over our minds,*" and "*They attained power by overthrowing the legal government.*" That certainly sounds logical enough, and yet it tells us nothing about how that control is occurring. "*Words have tremendous **control** over our minds?*" "*They attained **control** by overthrowing the legal government?*" What forces are acting upon your mind as you read this, or anything else for that matter? Emotion? Reason? Logic? All of the above, or none of the above? Which one is controlling you?

Likewise, what happens when a revolutionary overthrows a government? Did they simply kick some people out of a room? Did they have to announce they are in charge? Was it only a matter of raising a flag? Did they have to gain consensus of the newly controlled? Or did they just point a gun at everyone? If they were pointing guns, how did the holders of the guns decide who is in control of them and where those guns should be aimed? Who recognized this control and what made them do so to begin with? Was it by force? Philosophy? Was it the triumph of better ideas, or just better salesmanship? What legitimized that seizure of control and who is to say whether or not it is legitimate? Is it by consensus of the other "controllers" in the region? Or does legitimacy solely rely on the will—or the fear—of the people subject to the new control?

Likewise for *authority*. Authorized by whom, exactly? Through what mechanism does this authority flow? Was it ordained by God? Or does the authority derive from the will of the people? And if so, how do we know? What if the people were misled? Does that negate the power of the people wielding authority? Our current Congress is completely untrusted by 80% of voters,[3] yet they still retain authority with scarcely a second thought. Similar issues arise with *force*. Are we talking force of will? If so, how does that will get imposed on another? Is it because of a threat of physical violence? Is it fear? Hope? Is it aspiration, or is it driven by greed? Or is it the result of actual physical violence? Does the act of physical violence suffice to gain power? Or is there more to it than that? As I hope you are starting to understand, much like money, power is not a straightforward concept. Indeed, it is risky to assume that we know what power is, let alone make assumptions about who wields it and how or why they are able to do so. Those risks become far more dangerous when we assume that modern power structures in society are a natural phenomenon. They are not. They are an abstract construct, just like money.

As we will see in the later chapters on money, what follows will likely challenge a number of deeply held convictions. Ideas and concepts around power are rarely parsed out in detail beyond a select few academic fields.

For most, discussions around power too often rely on heuristic assumptions. Far too rarely do they go any deeper. With this in mind, if we are to unpack ideas around the alignment of incentives throughout the global political economy, it is imperative we give thought and weight to concepts beyond mere economics. I started the book with a discussion around power for a reason. It is a fine start to understand where money comes from and why it takes the form it does today. It is a great improvement to understand the structures that evolved to create and support that money in the first place. It is an error to assume that we know what power is without detailing how power exists. It is a much larger error to give scant weight to those dynamics while presuming good ideas and simple measures will suffice to overcome them.

To briefly recap, power is often interchangeably used to describe strength, control, authority and force. Much like money, a lot of ink has been spilled detailing power, power hierarchies, and power dynamics. The aim for this section is to break those concepts down to their constituent parts in a systematic fashion. To achieve this, we will start by tracing out the first component-part of power: Strength. Subsequent chapters will likewise be devoted to the sub-concepts of control, authority and force. The end goal of this endeavor is to provide a coherent understanding of power within the greater picture of modern global human interaction. It is also to reveal that, like money, the concept of power is much more complex and nuanced than what is commonly assumed.

CHAPTER TWO: STRENGTH

When you think about strength what first comes to your mind? A large, well-muscled man? A thin woman working two jobs while raising a child? A sturdy building? A military parade? Like power, strength has a number of connotations and invokes a number of different images. Oftentimes the mere mention of strength conjures a convenient bundle of all the images above together. Strength can then, and often does, become an all-encompassing character trait that loosely coalesces around combinations of attributes. Commonly lumped together are traits such as endurance, tenacity, perseverance, and the ability to use or project physical force. If you lump those traits together in a little heuristic[4] basket it becomes easier to see how we arrive at our nebulous understanding of strength. If we trace these notions back even further, the root of these concepts can arguably be found in Charles Darwin's work. "Survival of the fittest" forms the cornerstone of Darwin's theory of natural selection.[5] So ubiquitous is this phrase in modern parlance, it is dogmatically accepted. Its meaning is rarely questioned. As such, and more often than not, the phrase is bastardized to imply, "only the strongest survive." Contemporaneous in this bastardized understanding is the implicit acceptance of another common aphorism, "life is a competition for scarce resources." Expanding this understanding out, we arrive at the rough notion that only the strong can succeed in this deathly competition to acquire survival resources. Through that strength, it is presumed, survival is assured. This is all well and good, save for the minor inconvenience that none of it is correct. Let us dispense with some of the myths, starting with the *survival* portion of "survival of the fittest." I will defer to the inestimable Dr. Robert Sapolsky[6] here for the definition of evolution:

> *Evolution rests on three steps: (a) certain biological traits are inherited by genetic means; (b) mutations and gene recombination produce variation in those traits; some of those variants confer more "fitness" than others. Given those conditions, over time the frequency of more "fit" gene variants increases in a population.*

You might notice, and as Dr. Sapolsky goes on to say, survival is not a component here. Evolution is about reproduction, not survival. The better a species is able to pass on genes is what is meant by being the fittest. Which also means, fitness is not necessarily about being the strongest either. There are, in fact, inherited genetic traits that confer a greater ability to reproduce at the expense of the life span of the genetic donor. Life span, of course, being a fundamental characteristic of surviving. This function of increasing the odds of reproduction at the expense of survival is referred to as antagonistic pleiotropy.[7] The effect of which is demonstrably present in everything from salmon to primates. This leads us to another issue with misconceptions about survival of the fittest and selection. We often hear people speak of "alpha" males. In fact, a cottage industry[8] has veritably sprung up around the notion. In the lay understanding of an alpha-male, it is the strongest and most fearsome that gets the first crack at the food. Likewise, so the story goes, these alphas also get the first crack at mating. All this occurring while they stand atop a dominance hierarchy they seized through physical combat. Wolves are regularly paraded as the prototypical pack animal that adheres to this supposedly Darwinian hierarchy. The trouble is, and especially with wolves in particular, it is not true.[9] The biological problem with the idea of big, powerful alpha males dominating a pack comes down to variations of what we mean when we talk about selection. There is natural selection, which tends to favor traits that ensure genes are passed on. Traits like resistance to disease, better blood flow, stronger kidney function and the like. Then there is sexual selection, which tends to favor things like big horns, large muscles, or brilliant plumage.

The biological trouble arises from the fact that those traits can be at odds with each other. Big horns, or bulky muscles might win the affections of a female, but they are also metabolically costly. Those with a lower metabolic cost may well live on to reproduce for longer than their bulky sexual rivals. As Dr. Sapolsky asks,[10] "Which wins—transient but major reproductive success, or persistent but minor success?" As you may have guessed, the answer is, "It is complicated." The point here is not to retrace Biology 101. It is rather to demonstrate that commonly held assumptions about strength and its relationship to power do not necessarily hold true. The issue we run into occurs when discussions around power are unconsciously rooted in ideas that are not fundamentally sound to begin with. "Survival of the fittest" has a nice, easy ring to it. Within common parlance, it also conveys a false sense of how human beings achieve positions of authoritative strength.

Likewise, when combined with an idea that says, "life is a competition for scarce resources," we mistakenly paint a picture that is far too simple to convey the reality of human social dynamics. In the animal kingdom, hierarchies certainly form. Yet, those hierarchies are neither always, nor are they necessarily related to power dynamics. Worker ants are not relegated to the role of a slave serving a queen. Rather, when looked at scientifically, the reality is the ant colony is more akin to a single living organism. The mistake we made before this discovery[11] was to anthropomorphize the ant colony and assign familiar roles based on human experience. Similar for the wolves, where the mistake there was through observation of captive specimens,[12] rather than their wild counterparts. The point being, it is a tricky business to discuss strength in the abstract, such as when discussing a "balance of powers" between nation states. That power structure has a long history of development. Make no mistake, it was developed by man, flawed as he is. It certainly did not emerge as part of an inevitable evolutionary process in harmony with the laws of nature.

As noted at the outset and above, modern political power structures are not a natural phenomena. As we will see in the next section, much like money did not evolve from barter to currency to credit and debt, neither did the power structures we live under today evolve from brutish dominance hierarchies to representative democracies. Meaning, these power structures did not evolve from natural pecking orders inherent to our species. If they had, then power structures throughout humanity would all adhere to the same model. What we find are similarities in some cases, whereas in others we find complete divergence. The governance structures found in North American native tribes were incomprehensible to the Euro-Christian immigrants arriving on the Natives' shores. As the saying goes, history is written by the victorious. Without question, the victorious in the so-called "new world" of the Americas were very much the European Christians. But just because they won does not mean that the Natives did not successfully manage their political affairs prior to European influence. With that lengthy preface in mind, we can proceed apace with the discussion at hand. Ideally, that brief foundation will allow us to unpack and better understand where the notions of strength and power loosely outlined above take their root in Western society today. I hope it obvious why the focus is on the political economy of the Industrialized West. If it is not obvious, that focus simply stems from the observable reality that the global political economy is dominated by Western ideology and structures. Central to that ideology and those structures are the concepts of nations, nation-states, and national identities. Within those broad concepts lie the ideas of authority and sovereign control. Moreover, and as will be noted in the section on Money, the entirety of the modern global economic system is built upon the United States dollar, a Western construct to its core.

That said, a good starting question is, "Where does political strength in a nation come from?" The students of political science and political economy know at the outset that the origins of the idea of a *nation* are not easily discerned. Scholars trace the concept to a number of periods and for a number of reasons.[13] For the purposes of this discussion, we will affix a somewhat arbitrary starting line at the *Treaty of Westphalia*[14] in 1648. This document, which ended decades of conflict in Europe, was one of the first to clearly articulate ideas of territorial sovereignty[15] and the right of sovereign control. This is not to say those notions did not exist previously. This document rather provides a convenient starting point for the broader discussion. As we will see in the section on Money, if traced back far enough,[16] the dynasties that became the sovereigns of Europe by the age of the *Treaty of Westphalia*, were rooted in what could only be called protection rackets.[17] For a mostly accurate and clear modern example of medieval European power structures, one only need to look at Afghanistan today. After nearly 100 years of civil war and military occupations by the Russians and the Americans, the country has essentially been bombed into the early Middle Ages.[18] What arose in the vacuum created by all this conflict are locally powerful groups. Gangs, if you will. Their governance structures are nearly impenetrable to outsiders without verifiable familial connections. Within their spheres of influence, they impose their will on the local populace[19] through force of arms and sheer intimidation. Through that imposition of their will, they also enrich themselves through extortion. This is especially true against oppositional forces and non-familial, or other out-groups, such as those with different religious views. Once in control, these groups pick and choose the targets of extortion and to what extent. In-groups are heavily favored in those decisions, which leads to greater loyalty to the local gang. This helps distribute power down, as out-group members are at high-risk during any conflict or disagreement with an in-group member. Thus, on balance, localized inter-personal conflict diminishes, with out-group members suffering the most inequity.

In turn, the expectations upon social and personal conduct are heavily dependent upon complex code-based honor systems within both the in- and out-groups. Those honor systems are such a fundamental component to governance that violations of the honor code and challenges to personal or familial honor are potentially lethal affairs. Perceived or actual affronts to someone's honor can, and often will, lead to protracted conflict and blood-feuds that can span years. These conflicts are often to the benefit of the local gang leader,[20] who may actively encourage them to further cement in-group loyalty by "playing sides" and encouraging conflict with oppositional groups. Across the nation, these low-level conflicts between various local gang leaders led to sustained periods of economic decline and general chaos, especially in densely populated regions. In response to that general chaos and decline of social order, a group of these "warlords" banded together with local religious leaders and enacted a generally applicable code of conduct throughout the country. This code of conduct is based on Islamic law and tradition, and is strictly enforced. Violations of these codes of conduct are met with harsh punishments, up to and including public executions for things as trivial as theft or adultery. Through this fear mechanism, and other heavy-handed and repressive tactics, a semblance of order has generally been restored throughout the country. The unfortunate after effect is that this structure favors and incentivizes in-group abuses that must be continually checked by the coalition leaders. However, that is a delicate balance[21] for them to maintain. If their attempts to restrain behavior are perceived as dishonorable, or unfair, it invites challenges to the coalition leadership by violent upstarts who may be profiting handsomely from their extortive or other self-serving behavior. The need for these leaders to maintain this delicate balance provides ample opportunity for abuses by lower-order members against the populace to continue largely unabated. This tends to suppress overall economic activity and growth. Of course, this is a vicious circle, where general prosperity remains low and depressed. Meanwhile the favored few are able to exist in relative comfort. The incentive structure is heavily weighted towards general stability,[22] with only the most basic needs being met at a subsistence level. From the perspective of the coalition leaders, however, it is much superior to unchecked internecine violence, chaos, and ultimately, mass starvation and death.

Put another way—from their perspective—they are doing the best they can with what they have available. After over 100 years of civil war, invasion, and occupation, what they have available is very little. What has worked best for them so far are:

1. Strict behavioral edicts enforced through violence,
2. The imposition of strict religious code adherence; and
3. Prevailing upon and exulting honor codes to better regulate interactions between previously warring factions.

With all that in mind, if we trace back the history of England—our legal and structural forbearer—to the time before the *Magna Carta*,[23] we will see some very familiar themes. After hundreds of years of invasion and civil conflict,[24] locally powerful groups, reliant upon strict admission criteria based on familial ties, rose up and imposed their will upon the local populace through violence. Through that imposition of their will, they also enriched themselves through extortion, especially against out-group members. This led to a lot of conflict among the locally powerful groups, which resulted in chaos and decline of social order, especially in densely populated areas. In response to that general chaos and decline of social order, a group of warlords banded together with local religious leaders and enacted a general code of conduct throughout the country. This code of conduct was based on Christian law and tradition, and was strictly enforced. Violations of these codes of conduct were met with harsh punishments, up to and including public executions for things as trivial as theft or adultery. In fact, if one traces the roots of criminal law, there were only nine felony crimes[25] under English common law: Murder, robbery, manslaughter, rape, sodomy, larceny, arson, mayhem and burglary. All were punishable by death via public execution.

As an interesting side note, the term "common law" refers to the idea that prohibitions against things like theft or robbery did not need to be spelled out. Everyone, throughout the realm, knew that stealing (for instance) was wrong. Therefore, it was common throughout the realm to prohibit the act of theft. That note aside, governance and control of behavior in the early Middle Ages in England was, much like Afghanistan today, dictated by an honor-code system. For a case on point, small English settlements were often referred to as shires. Each shire would have a handful of families living within it. Each family was liable for the behavior of every other member of their family. If your uncle stole something, you, as a nephew or niece, were responsible for that theft just as much as if you had stolen something yourself. Thus, every interaction you had with your neighbor implicated the family honor. If your uncle was caught stealing, the code demanded the family preserve their honor by harshly and publicly punishing their own kin. Carrying this socio-behavioral regulation methodology further out, each shire would elect a shire member to be a representative[26] for everyone in the shire. This person was called the Reeve and they were sent to mediate disputes between the shires and to coordinate trade rules and other matters. The Reeve was also responsible for mediating disputes within the shire and could compel punishments for violations of the honor code, as well as make determinations of whether or not honor had been satisfied. In this regard, these Shire Reeves were empowered to be part legal representative and spokesperson, part police officer, and part judge. If you trace it out a bit, this early English power structure and authority still exists in the United States to this day. In fact, we still call them Shire Reeves, though a few hundred years have truncated the term to Sheriff.

The Shire Reeve was a noteworthy moment in early English society. Within that structure we start to see the beginnings of the English form of representative government. A representative government influenced heavily from the Greek and Roman traditions. It also marks a milestone for the softening of the highly punitive and strict behavioral controls that preceded the Shire Reeves. The broad argument here being, if you give the Afghans enough time, they will probably come up with something similar, if they have not already. Nevertheless, as these English honor systems refined and ossified, they gave rise to the incredibly complex honor system still in use by the English nobility today. Title and rank being bestowed for great acts, sacrifices, bravery, and so on. The *Magna Carta* itself was another milestone achievement, as with its introduction we can see the first formal steps to codifying a set of principles intended to check and prevent the abuses by local warlords that are readily observable in modern day Afghanistan. Ultimately, the formation of these systems, diverse by hundreds of years, reveal striking parallels. Like Afghanistan today, the early English system of governance formed in much the same way and with the same core outcomes. Reflect for a moment that, in early England, governance and behavioral regulation was achieved via:

1. Strict behavioral edicts enforced through violence,
2. The imposition of strict religious code adherence; and
3. By prevailing upon and exulting honor codes to better regulate interactions between previously warring factions.

It is important to note here, however, that when discussing power structures such as the one in its infancy in Afghanistan, and the much more developed version in the United States, the foundation of both resides in behavioral edicts enforced through punitive violence and expectations of code-based honorable conduct. Oddly enough, there is another strikingly similar, and parallel, power structure that exists in the United States. It arose within the U.S. prison system. As author David Skarbek lays out in great detail in his book, *The Social Order of the Underworld*,[27] the prison population expanded dramatically during the 1970s. Prior to that expansion, prison social dynamics were loosely governed by an honor system known as the convict code.[28] Strict adherence to the code provided a quantifiable measure of honorable conduct. The reputation gained by adherence to this code was valuable to the holder. Their honorable reputation was the yardstick by which they would be measured when they found themselves housed in a new, or unfamiliar prison. However, once prison populations exceeded the convicts' ability to rely on honorable reputations to maintain order, mass violence and chaos ensued. In response to this violence and chaos, prison gangs began coalescing around local leaders. These leaders, much like their ancient counterparts in England, and their contemporary counterparts in Afghanistan today, are all particularly effective at both violence and persuasion. Once these prisoners coalesced around selected leadership, they enacted forms of governance not dissimilar to the ones seen in medieval English shires.

Much like the families of the shire, each gang is responsible for the conduct of all other gang members. If a member cheats or steals from a rival gang, the expectation of the rival gang is that the offender will be held to account by his own gang members. Depending on the severity of the honor violation, gang leadership may impose punishments that range from forcing an apology, to beating their own misbehaving member, all the way through to killing them. In this way, the gangs, by agreement, use the collective honor of the gang to enforce social norms and prevent wide-scale violence and chaos. If the other gangs' honor is not satisfied, then violence between the groups may erupt. Thus the honor code among the groups is highly prized and earnestly defended. I point this all out here because it demonstrates that, among these disparate groups, a very similar pattern emerges. Authoritative rule-making power, in all these cases, is *granted* after leaders emerge that seek to reign in widespread chaos and disorder. Because the chaos and disorder is so great, in all these cases, it requires harsh means and strict edicts to control it. In-group favoritism helps cement the loyalty of lesser members of the leadership group. But this also makes that leadership tenuous and difficult to manage, as missteps may have fatal consequences. Thus, abuses by lower level leaders may be difficult or impossible to reign in, unless they are egregious. In an effort to combat this problem, all of these structures exalt and foment honor codes. They demand honorable conduct of both the group and the individual. To ensure honorable conduct is adhered to, these honor systems all require in-group self-policing and appeals to out-group consensus on the sufficiency of honor satisfaction.

However, when it comes to achieving and wielding leadership roles in these disparate settings, what we see is not simply a matter of using violence. Subservience to the power structure is achieved through deference to leadership that arises in response to chaos and unchecked violence. Requisite to this idea, though, is the need for collaboration, compromise, all while ensuring incentive alignment among the in-group members, and treaty-like conditions with out- or oppositional group members. Strong leaders do not simply appear, beat the others down and then lord over them. Rather, they are highly effective at achieving consensus among brutes. This is no small task. The reasons hierarchical authoritative rule-making leadership becomes recognized are certainly buttressed by the capacity for those leaders to wield violence. But that is not the foundation of their authority. Wisdom, political savvy, risk management, and competency all play a crucial role. Maintaining the position requires even more finesse[29] and it rarely lasts. Given these constraints, it is no surprise the institutions that evolve from these lines of authoritative leadership can take decades or even centuries to develop. Such is the case with the English through the Middle- to High-Middle Ages. That system of authoritative rule-making—replete with courts, judges, parliamentary rule, representative participation, and written law— was entirely derived and refined from the rather brutish conditions outlined above. Likewise, how those systems are administered and maintained today reveal their very clear lineage to strict behavioral edicts enforced by violence and appeals to code-based honor adherence. Nevertheless, it is important to stress the fact that the origins of authoritative leadership were achieved through rough consensus and cooperation. They were most certainly not created solely through the use of coercion or coercive force. Moreover, in all the cases described above, those lines of authoritative leadership were reactionary responses to chaos and disorder. Which is to further say, they did not necessarily evolve from base impulses to seize power. Indeed, the entirety of this line of reasoning comports quite well with the Hobbesian view of "War of all against all"[30] modified only in the regard that the war of all against all eventually becomes intolerable and invites some level of cooperation. Nevertheless, it rather appears to be the case that leadership and rule-making authority emerges in response to chaotic attempts to seize control of resources or territory. Moreover, after these rule-makers end up in power, they often come to realize their position is tenuous and entails great personal risk.

Understanding this as we do now, it is appropriate to trace back to the start of the discussion. Recall, local leadership begins with verifiable close familial connections. Those groups then tend to impose their will upon the local populace through violence and extortion. Conflict between similarly situated neighbor groups eventually declines into chaos and disorder. In response, these groups coalesce into a rough governing body, usually with the assistance of recognized religious leadership. From there, authoritative rule-making leadership gradually expands outwards to more distant groups that roughly share language and religious traditions. At some point, however, that expansion runs into a language or religious diversity that is hostile or resistant to incorporation. It is roughly at that boundary line of religious and language divergence that one groups' authoritative rule-making leadership ends and another groups' authoritative rule-making leadership begins. What the *Treaty of Westphalia* did was to draw those lines on a map and say that, once and for all, everyone behind one line belonged to one group, while everyone on the other side of the line belonged to another. Thus, the modern nation was born.

I hope it obvious that much history is skimmed here. A definitive description of the historical record laid out above would require volumes. The radiation of concepts around governance, empire, philosophy and religion are deeply interwoven on the European continent. From the ancient traditions of the Greeks and Romans, to the proliferation and material blending of Christianity with Celtic paganism, the conflict with the Eastern empires and Islam, the Frankish conversion, the Christian divide and thousands of other political and migrational shifts on and around the continent, the history of European power structures is immensely complex. Thus, the effort here is not to ensure some form of historical fidelity. It is rather to hypothesize and, ideally, sum some of the core historical underpinnings of the modern power hierarchy, especially as it applies to the West broadly, and the United States specifically. More importantly, it is to point out that the power structure we live in today was born of very practical considerations. When discussing strength as a prelude to achieving or maintaining political power in the Western world, the broad takeaway should be that rule-making leadership very generally tends to arise in response to chaos and disorder. But that chaos and disorder very often originates from the unchecked use of semi- to formally organized violence by an in-group for that in-groups' benefit.

Perhaps counterintuitively then, often it is these same groups that end up creating a new power structure with rule-making authority. Which leads to the rather reasonable conclusion that sustained chaos and disorder is simply an intolerable condition, even among those who create and exploit that chaos to begin with. To trace the process out a bit, in a power vacuum, the short-term incentive is to gang-up and competitively seize as many resources as possible. That short-term resource competition leads to increasingly negative and costly outcomes. This then paves the way for cooperation and coordination under a unified ruleset. Within a unified ruleset, rudimentary institutions such as dispute resolution, collective defense, and representative bodies are given latitude to form. These institutions gradually codify hierarchical roles in the new power structure. Over time, those institutions grow in complexity and sophistication. Through this growth, the power structure naturally becomes increasingly bureaucratic and leviathan.[31] If the power structure is adaptable, it tends to survive for extended periods. If not, the power structure collapses and the process begins anew.

The point being, when discussing the creation of state power or state authority, what we find is the process is not simply born of brute force. It is often rather a response to brute force that requires some level of cooperative engagement between competing factions. In some cases, like the early Greek cities, geography played a central role in the diffusion of power among the cities, with cooperative trade becoming preeminent over conflict. Material abundance in the rich Mediterranean region certainly helped. Nevertheless, whether that cooperation proves sustainable tends to vary widely, and is often dependent upon the wisdom and foresight of the brutes that decided to work together in the first place. Since brutes often lack wisdom and foresight, successful implementations tend to favor power structures that interweave religious doctrine as a cornerstone for their rulesets. Moreover, a divine purpose,[32] and indeed, a divine blessing can serve as a useful motivator to achieve broader consensus among the governed. A religious foundation also serves to limit or remove perceptions of self-motivation among those seeking to solidify or increase their rule-making and rule-enforcing authority. But at its juicy core, deference to authoritative rule-making is generally granted because the rule-makers are providing relief from chaos and disorder. Meaning, for the purposes of the discussion at hand, and only broadly speaking, political strength is often first achieved by bringing order to chaos.

With that said, I hope it obvious that political power dynamics are not that simplistic. The aim here is to establish a very generalized baseline for where concepts of authoritative rule-making and deference to that action likely emerge. Among all living beings, a very clear through-line is the quest to reduce or mitigate entropic forces. Biological systems, by their very nature, are continually seeking homeostasis.[33] Biological entropy is death. Thus, it is a natural point of origin for hierarchically cooperative species, such as humans, to elevate group members that are most effective at reducing perceived or actual chaos. Whether it is an office worker that "takes charge" in an emergency, a fireman at the scene of an accident, or indeed, a local gang-leader holding a peace summit, people naturally gravitate towards those that most effectively keep chaos at bay. Thus, while political strength may shift and turn, at its core, respect for that strength is principally reliant upon the perceived or actual ability to achieve, or maintain order.

CHAPTER THREE: CONTROL

As we learned in the last chapter, a governed population will readily defer to actual or perceived strength when confronted with sustained chaos and disorder. That the chaos and disorder was often brought about by the very same (or closely related) people that end up stopping the chaos and disorder is rarely questioned. To quote Littlefinger from Game of Thrones,³⁴ "Chaos is a ladder." Once chaos and disorder are set in motion, relief from those forces is a powerful and soothing tonic. This, however, has a number of implications for achieving political strength in a given nation. In a legitimate power vacuum, like one might find in a failed state, when a group arises that brings order to chaos, the strength is usually self-evident, locally sourced, and generally well-tolerated—if not outright embraced. However, as the US Central Intelligence Agency (CIA) has figured out through the years, you do not always have to wait for chaos to emerge. Sometimes you can just go and create some chaos³⁵ of your own. Then, if you are really clever, you can shape internal power structures to better suit your needs or desires.³⁶ Oftentimes, and under those external conditions, the need for that strength display will be less self-evident. Creating conditions to reinforce the "need" for intervention will also require more convincing, such as through propaganda campaigns, to get the point across.

Regardless, whether your demonstration of strength was in response to internally created chaos and disorder, or if the chaos and disorder was externally manifested, it is the perceived or actual ability to bring order that generally grants you authority to make rules. Once you have rules in place, then your next step is to figure out how to enforce them. One of the reasons Western societies tend to gravitate towards honor systems is they offer an easy conduit for self-imposed behavioral regulation. When your reputation is exulted and given high status then protecting that reputation becomes very important. This is as true for inmates in US prison gangs today as it was for medieval soldiers in an English kings' army. This begs the question though, "where does this highly prized sense of honor come from?" The short answer is: fairness. The trickier part of that answer revolves around whether or not expectations of fairness arise from competition, or through cooperation. To skip ahead for a moment to the discussion about money, what we will see is that most trade between trusted parties is based on cooperation and mutual benefit. For an easy illustration of this process unfolding in chaotic situations, just imagine if you and your best friend were stranded in the Canadian wilderness during the winter.

If your friend is adept at making bows and arrows and you are adept at making clothes, it is beneficial for each of you to supply the other. Obviously, if you have a bow and arrow and your friend is not freezing to death, the likelihood of you both surviving goes up. This is the essence of group cooperation among humans at the survival threshold. Similar behavior is readily apparent[37] among wolves, wild dogs, herd animals, primates—even amoebas. Cooperation is a cornerstone of pack animal survival. Yes, individually, they may squabble about things from time to time. Often those squabbles revolve around issues of fairness, like what might unfold if someone snatches fruit out of your hand. But at the group level what we find in these dynamics is cooperation generally trumps individual desires. The group acts in concert to acquire resources, such as food, and bands together for protection. General cooperation is the through-line. Pack leaders guide the gathering of resources and expend a fair amount of effort to ensure that the weaker members, such as the elderly and the young, are provided for. Group members that egregiously violate the "rules" of fairness are often ostracized, or put in check by the pack leadership.

Put in human terms, as Justice Oliver Wendall Holmes once said, "Even a dog distinguishes between being stumbled over and being kicked." Where this starts to go wrong is when competition is introduced. While it is a common assertion that warfare among humans is a natural condition for mankind, this is not necessarily true. As Dr. Sapolsky points out,[38] among pure hunter-gatherer tribes, wide-scale conflict, such as warfare, is a rare occurrence. An example he gives is of a group hunting in a particular location. If they came across another group hunting in the same area, they would not fight over who got to hunt. They would simply go somewhere else. One of the few archeologically documented instances of wide-scale conflict among "pure" hunter-gatherer groups[39] was in a location that had abundant fish and wildlife. What seems to have occurred was a group set up shop and then tried to exclude others from fishing there. Or in the alternative, another group simply decided to muscle in and kick the first group out. Meaning, once competition is introduced through exclusion or attempted exclusion from a shared resource, group fighting becomes more likely. The larger trouble is the historical record is spotty at best. There may have been any number of large scale conflicts that are lost to history. Thus, it may be true that warfare is a natural condition for humans. It just does not seem likely based on what we have been able to determine. What we can glean from that spotty historical record rather supports the idea that, prior to the adoption of agriculture and animal husbandry, nomadic hunter-gatherers tended to be relatively cooperative and peaceful. This is not to say they existed without violence at all. It is rather to say that organized campaigns of violence like what the European, or the Middle-Eastern and Eastern dynasties routinely engaged in, tended to be rare.

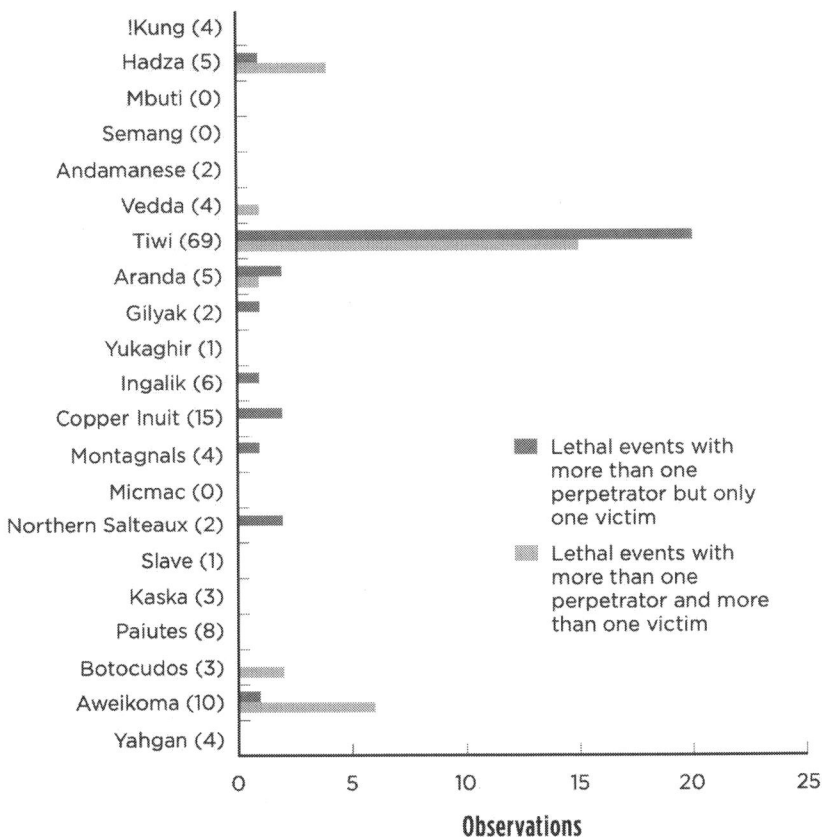

As alluded to above, it seems that once agriculture and animal husbandry became a regular facet of collective life, organized campaigns of violence followed. This makes intuitive sense, especially when considering the otherwise cooperative nature of the human species when living at the survival threshold. Again, what is implicated traces back to the idea of fairness. Who should benefit from the work of raising and harvesting a crop when the yield of the crop far exceeds your ability to consume it? Likewise, if I am hungry, but you have captured all the goats, what gives you the right to keep all the goats from me? Meaning, where it all goes a bit askew is when resources are gathered in mass, like from a harvest, or with herds of domesticated animals. Whether by ingenuity or by force, once the notion of "haves" and "have-nots" sets in, it gets tricky to figure out what is fair. It gets even trickier when a bunch of strongmen come around and lay claim to all the land and everything on it.

As roughed out in the previous chapter, what this type of competition for resources eventually devolves into is chaos. The thing with farming or raising animals is those actions also require some level of cooperation, depending on the crop or the animal. This is evidenced by the philosophies and the gods that emerge from those societies. For instance, growing rice requires significant labor input and the development of a fair amount of agricultural technology[40] to accomplish at scale. To successfully feed people with rice requires all hands be on deck. Moreover, those technology developments radiated outwards and had a profound effect on agricultural practices in regions where other crops, such as millet and soybeans, were raised. By contrast, cultivating wheat, which is a grass grown in soil, requires less labor by comparison. Likewise, shepherds can corral a fair number of animals by leveraging and manipulating herd animal behavior. However, if fish are your primary animal protein source, it again requires some level of cooperation and group effort to harvest a large number of fish, especially from the sea.

Thus, we find loose correlations between societies that grew and developed agricultural technology around rice and fishing, like those found in the coastal regions of the East, creating philosophies and gods that orient around collectivism.[41] Meanwhile, societies that grew around wheat and domesticated feed animals tended to create philosophies and polytheistic[42] gods that oriented and vacillated between collectivism and individualism.[43] By far the most individualistic philosophies and staunch monotheism[44] arose among the nomadic, shephardic pastoralists,[45] such as the Akkadians, Aramaeans, Arabs and Hebrews. Tracing back further, societies that grew around hunting and gathering, such as the Celts in Europe and the American Natives in the far West, tended to create philosophies and gods that oriented around natural forces.[46] Within this, they elevated neither individualism, nor collectivism and rather placed mankind squarely within the greater natural system. Of course, bits and bobs of each find their way into all of them. But generally speaking, the major through-lines are self-evident in the major religious doctrines of each: the duality and balance of Taoism and Buddhism in the East; the righteousness and vengeance of Islam and Christianity in the Middle-East (and eventually into the West); and the natural order animism[47] of animal and weather spirits among the hunter-gatherers.

What is noteworthy here is that notions of honor are prevalent throughout these divergent systems. How that honor is expressed and exalted,[48] however, can be markedly different. If you are a member of a society that relies on wide-scale collective effort to ensure everyone is fed, then honor will naturally coalesce around actions that reify inclusion in the group effort. By contrast, if you are a member of a society that relies on individual effort, then honor will naturally coalesce around actions that reify each individual's effort. Turned to the negative and boiled down to a simplistic frame, if you bonk someone on the head and steal their rice or their wheat, it is, ultimately, an affront against the effort of the group. Yes that theft affects the individual as well. But if that is allowed to happen often enough, the affront will impinge upon and impede the group effort, which negatively impacts all.

By contrast, if you bonk someone on the head and steal their goat, it is an affront against the effort of the person tending the goat. In that case the goat only changes ownership. But any larger group associated with the shepherd remains in the same position relative to the goat pre- and post-theft. The group did not put any labor or effort into the goat to begin with, only the shepherd did. Meaning, unlike the farmers, they had no claim on the goat regardless of the current owner. Group censure and ostracization are natural and predictable responses in the first instance of collective affront. Vengeance and retribution are natural and predictable responses in the second instance of individual affront. For a modern analogy, imagine you are on a baseball team and you discover one of your teammates is cheating while betting against the team. How do you imagine the team will respond? Change the scenario a bit and imagine instead that you are a boxer and you discover your opponent was cheating by loading his gloves with metal. How would you respond then? In the first instance, it can be reasonably anticipated that—absent a governing body—the team would shun or remove the offending member. In the second instance, it can again be reasonably anticipated that—absent a governing body—your next match will likely find you sporting some metal (or worse) of your own.

Thus it is unsurprising that honorable conduct in the East is generally predicated[49] upon doing actions that do not dishonor the collective or the group, with shame and shaming practices featuring prominently. Returning briefly to the earlier part of the discussion on the limits of sovereign control, it is at the rough boundary of ancient Mongolia and China where the Chinese influence waned. Recall territorial influence tends to diminish in regions where language and religious practice diverge. It is on the ancient steppes of Central Asia where those languages and religious practices did in fact diverge. Early religious practice in Central Asia generally revolved around Tengrism,[50] which was based on shamanistic animism. Given the harsh and sparse geographic conditions, it makes sense that the gods chosen would reflect this. It is also no coincidence that the nomadic Mongols ended up with power structures that featured tribal rule and internecine conflict.

In the West, and similar to the Mongols, honorable conduct is generally predicated upon doing actions that do not dishonor yourself or another individual, with vengeance and retribution featuring prominently. This is not to say that honor exclusively orients between the collective and the individual in these spheres. It is rather to delineate the relative weighting of incentives within each respective backdrop. Expanding the collectivist versus individualist orientation outward a bit, it has been argued that military advances on the European continent were not simply a product of military competition. Instead, as the theory goes,[51] it was specifically tournament-style competition among individual states that led to rapid advancements in military tactics. By contrast, the dynasties of the East, such as in Japan and China, enjoyed similar technological instruments, like siege weapons and firearms, as their European counterparts. But it was the lack of tournament-style military competition that caused those empires to lose ground to European tactics.

What prevented tournament-style competition in the East? Unification of competing interests into singular territories—Japan and China respectively. On the European continent it was quite the opposite, as was demonstrated earlier with the *Treaty of Westphalia*. What we see with that document is the codification of individual sovereign rule over specific territories, rather than combining into a singular European nation like the Chinese and Japanese did. Perhaps more importantly to the discussion at hand, it is noteworthy here, that prior to the wide adoption of Christianity in the Roman controlled regions, Rome itself was not identically, but rather similarly, structured like the great empires of the East. The interesting thing is, the Romans were also never able to fully hold the areas of Northern

Europe, with their furthest forays ending in Southern England.[52] But the Roman presence and incursion into Germania and England paved the way for the introduction of Christianity into the region. Thus what we arguably have here is twofold: A gradual spread of agricultural practices that predates the Roman empire, coupled with an injection of an individualistic, vengeance based philosophy and religious practice into the broader areas of tribally dominated Northern Europe.

Recall the earlier question of where a sense of honor originates. The supplied answer was fairness. The caveat to that answer was whether notions of fairness are rooted in competition or cooperation. Christianity in particular traces its philosophical origins to the nomadic shepherding practices of the Levant.[53] Animal husbandry in the Levant primarily revolved around herding sheep and goats. As noted above, these practices tend to favor individual or small group effort. A hopefully obvious corollary to this is, an individual or small group is more vulnerable to robbery. One good bonk on the head and a thief can make off with all your goats. Once bonked and stolen from, a natural impulse is to try to find the assailant and retrieve the ill-gotten herd. If you find them and they do not want to give the goats back, then you will probably have a fight on your hands. But what if you get it wrong? What if the person you find has a herd that looks like your goats, but they are not? What happens if you bonk him on the head and take his goats by mistake? Chances are good he will come looking for you as well.

Take it a step further. What if you really do find the guy that stole your goats, but when you try to get them back, he bonks you on the head again. What if he is so good at bonking you on the head that you physically cannot take your goats back? Are they his goats now? Now that you are competing for ownership of the goats, what rules govern in the absence of rulers? This is the essence of where honor systems based on competition for resources are created. With a little imagination, it is not difficult to trace out how these competitive honor systems develop into notions of fighting fair, taking an eye for an eye, or not preying upon the weak. One of the big problems that arises from competition, however, is that competition itself has been shown to lead to unethical behavior,[54] especially among the winners. Coupled with the incentives for competitors to test the limits and push the boundaries of rules, it is not hard to see how competition for resources can quickly develop into a system of "might makes right" with only the thinnest of justifications. Thus, if you find yourself at the top of a dominance hierarchy, in control of vast resources, and are now ethically compromised from the effort, it kind of makes sense when you become a despot. Moreover, it makes it very easy to take the next logical leap, which is that you arrived at the top of this dominance hierarchy because of your superiority in competition for scarce resources. That the resources were scarce because of the type of crops your ancestors grew, the animals they chose to domesticate, and the people that tried to control them is immaterial.

This is especially true if you are many generations removed from your hunter-gatherer ancestors. The idea that food and other resources are scarce can begin to seem very natural when your food and other resources are concentrated and the means of production is "owned." Meaning, this scarcity mindset is magnified when the only visible source of those resources are being piled up in someone's barn and you cannot get to it. Mind you, resources can also be scarce due to poor weather, bad yields, or disease. But once someone starts laying claim to parcels of land—or indeed people—the scarcity is more often than not going to be driven by theft or capture, and subsequent mismanagement of those resources. As alluded to above, the larger problem being, all of this is easily mistaken for the "natural order" of things. One only need look at a wolf tearing at a deer to presume that violent competition for life sustaining resources is the way of nature. What it fails to account for is the larger biological system, which is actually trending towards homeostasis when that wolf consumes that deer.

None of this is to say that animal husbandry and farming are the cause of all woe and misery. Indeed, a few hundred years of rapidly expanding human populations very clearly demonstrates that both have led to enormous advancements for humans to thrive as a species. Moreover, if eight billion people were to suddenly attempt to hunt and gather their way to sustenance, the planet would be stripped bare in a matter of months. Agriculture and animal husbandry are essential for continued human existence at the scale of the current global population. The purpose of this exercise is to demonstrate that the Euro-Christian philosophical worldview that underpins the current global economic order is built on very problematic foundations. Indeed, given the unique combination of agricultural practice coinciding with Christian worship in Europe just as Roman influence waned, it begins to make sense why the Europeans ended up with the type of governance structures that led to tournament style competition for land and resource capture.

Meaning, despite the fact that the Europeans engaged in many of the same farming and animal husbandry practices found among the Romans or the Eastern Dynasties, their particular social and governance structures were guided and built upon Christian based pastoralist world views and philosophies. With that notion put forth, it must be stressed that this all represents a gross simplification of the development of human behavioral control dynamics through history. As mentioned previously, none of this is intended to deliver a granular level of historical fidelity. The primary purpose of this exercise is to provide a deeper insight into the origin of these systems. It is also an attempt to break the reader free from the confines of often dogmatically accepted axioms regarding the nature of power, power dynamics and power hierarchies in modern society. It requires substantial effort to parse out a historical narrative free from the bias of the historians. Not to mention, much of Western historical study revolves around the history of the West. But human societies span the globe and have encompassed a rich tapestry of philosophies and cultural approaches to civilization.

The larger error is the assumption that Western culture is somehow "better" simply because it ended up becoming the primary ideological system governing trade and international relations modernly. As noted above, there are inherent problems with Western culture. Not least of those problems are the fact that it is largely predicated upon competitive resource capture and allocation. As I have attempted to demonstrate thus far, humans are not necessarily, nor naturally inclined to compete for resources. In fact, anthropologically speaking, we are rather more oriented and inclined towards cooperation. Traced back far enough, we can observe where modern notions of competitive markets and adversarial rule making systems emerge. Arguably, the same reasons those systems became pre-eminent are the same reasons they are net-destructive in their implementation. Much conflict, suffering, and death has been wrought upon mankind under these economic and social philosophies. One only need look to the so-called "primitive" tribes that still exist today, and indeed those that were our ancient forebears, to understand that humans are quite capable of existing harmoniously with each other and with nature. Yet this also does a disservice to the incredible human innovations and achievements that have come from a competition based system. Art, literature, music, harnessing electricity, instant global communication, peering deeply into the cosmos, heavier than air flight, and even the ability to leave the planet itself are nothing short of amazing. Moreover, we have enabled billions of people to exist on Earth simultaneously, expanding our collective consciousness far beyond anything our ancient forebears could have imagined. Nevertheless, it is fruitless to imagine "what if" and far more productive to address "what is." And "what is" currently can be reasonably portrayed as:

1. A global system
2. Based on competitive resource capture
3. With rule-based governing structures rooted in pastoralist traditions
4. That are created in response to the conflict and chaos that emerges from unregulated competitive resource capture.

While this may sound trite, it has profound implications. What it suggests, and what I argue here is, that any competitive system inevitably devolves into a rule-based control system. But if those rules are predicated upon philosophies deeply rooted in individualism, then tournament style competition becomes more likely. Nevertheless, in the broader context of the chapter at hand, it appears safe to say that top-down control uniformly manifests within competitive systems and less so among cooperative ones. A simple thought experiment should suffice to describe why this is so. Imagine there are two groups of ten people each. Each group has a two-year old child, a person with no legs, a frail eighty-year old man, a frail eighty-year old woman, and the other six are healthy, fit young men. One group only speaks German. The other group only speaks Hindi. Between them is a large, steep, rugged hill. On top of the hill is a large platform.

> Scenario #1: Whichever group has more of their people standing on the platform at the end of an hour will win $5 million each.

Scenario #2: Both groups must be standing on the platform at the end of an hour to win $5 million each.

There are no rules and no consequences except for losing. What comes next? Depending on how motivating $5 million is, Scenario #1 can get ugly fast. It would be chaos, would it not? Provided these people are not absolute psychopaths, chances are good they are going to try and come up with some rules of their own, even in the absence of "official" rules. Those rules might be based on fairness, honor, practicality, self-serving greed, or manipulation. Or perhaps they will just resort to brutality. Even if they do resort to brutality, provided they are evenly matched, sooner or later that is probably going to incur costs that far exceed the value of the $5 million. At which point the survivors will probably come up with some rules anyway. But what of Scenario #2? What would their structure be? Would they even need one? Maybe. Maybe not.

One need not even construct a hypothetical this complex. To simplify the point even further, just imagine an American football game. Keep the objectives the same, keep and hold a ball and cross a line at the end of a field, but take away all the rules. How long do you think it will take before each team is yelling and screaming at the other, or even openly fighting? In this American football example, the resource to capture is a fiction. Crossing a line at the end of a field while maintaining possession of a ball. The only value in the exercise is the competition itself. And yet, just that fictional resource capture game is enough, by itself, to require intricate and detailed rulesets to ensure fairness. Even with those rules in place conflicts are frequent. Disputes over those rules and gamesmanship often lead to physical violence on the field of play. In fact, there exists a rather fascinating virtual example of similar power dynamics as described here unfolding in cyberspace.

The massively multiplayer online role playing (MMORPG) game EVE Online is set in a fictional, futuristic space-faring universe. The game world favors and incentivizes resource capture and exploitation. The field of play is divided into three sections. Two large sections feature computer based non-player characters (NPCs) that provide security assistance to human players. The largest section has no non-player originated security interventions at all (Nullsec). What is fascinating about the game is what happens in Nullsec follows a very similar political power development dynamic[55] as outlined above. Early in the game gangs formed. They engaged in piracy and extortion. Conditions were chaotic. Groups started banding together, which led to more conflict between groups. Those groups eventually coalesced into major, regional factions that now control most of Nullsec. They enforce rules of travel and transit, levy taxes, and even wield policing authority within their respective spheres of influence. These major coalitions have even engaged in outright warfare with massive campaigns featuring tens of thousands of combatants that resulted in estimated real world losses of over $350,000.[56]

What we can glean from both the historical and modern examples provided thus far is a relatively straightforward thesis: Unregulated competition for resources leads to fighting and chaos. Fighting and chaos leads to the formation of powerful coalitions with hierarchical leadership. Hierarchical leadership leads to the formation of control structures that feature formalized rules for behavioral regulation within the sphere of coalition influence. But the formation of powerful social coalitions does not tamper or diminish the overarching competition for resource capture. The powerful social coalitions only increase the costs of acquiring more resources from the coalition's respective competition, e.g., group fights become increasingly complex wars. This dynamic is evident in U.S. prisons, modern Afghanistan, medieval England, ancient Rome and, indeed, the fictional galactic universe of EVE Online. The late-stage, nuclear-armed, mutual-assured-destruction, several thousand-year development of this very same power dynamic is on full display today. It underpins the entire global political economy.

CHAPTER FOUR: AUTHORITY

To briefly sum what we have covered, the modern concepts of political strength and control are born from unregulated competition to capture resources. In turn, agriculture and animal husbandry, driven by individualistic philosophies, tends to foster the environment for tournament style competitive resource capture. Within that game, political strength arises by bringing order to the chaos that occurs from the unregulated competition for resources. Once political strength has been demonstrated, the type of control wielded roughly correlates to the amount, or lack of, collective effort that is required to create the resources the politically strong seek to capture. More collective effort generally results in group-oriented control systems and social rules. More individual effort generally results in individually-oriented control systems and social rules. In both instances, honor plays a crucial role in group behavioral regulation. Collectivist honor systems tend to orient around shaming and ostracization in response to norm violations. By contrast, individualist honor systems tend to orient around vengeance and retribution in response to norm violations. In both cases, these honor-based behavioral regulation systems, given enough time and latitude, will generally congeal to become formal written rules.

It must be said here, prior to the advent of agriculture and animal husbandry—about 12,000 years ago[57]—pure hunter-gatherers very clearly did not create these types of social structures and rule-based systems to govern their social interactions. Generally speaking, absent competition, strength and control are expressed very differently in pre-agricultural human societies. Cooperative effort is seemingly the optimal state for human beings to interact with each other. Even in competitive situations, the better groups are able to coordinate and cooperate, the more successful they will generally be. Thus, it is an error to presume that competition is the natural state of human existence.

However, once competition is introduced into human interaction, the incentive shifts to use ingenuity and inventiveness to maximize competitive advantage. By definition, [taking advantage](#)[58] can mean to make good use of something. But it can also mean to take, use or expect something unfairly. If you are in a society that highly prizes individual honor, it becomes very easy to see how advantage seeking can be construed as dishonorable. And if, as is posited here, honor violations in individualist societies invoke vengeance and retribution, then it is perfectly understandable when conflict and violence naturally flow from unregulated competition. Thus, one of the major features of bringing order to chaos in competitive resource capture societies is to create rules to ensure fairness in competition. The question then becomes, who is allowed to enforce the rules once you create them? As we saw in the last chapter, those who take power are the ones that get to create the rules. But as anyone who has ever played sports knows, there has to be a neutral party to enforce the rules. Otherwise, the perception of fairness evaporates and chaos will resume. This is where the concept of authority is rooted.

Put plainly, being in authority means you are the one that decides what is right and wrong. As we saw earlier with the development of the English common law, there were rules of behavior that were generally recognized as self-evident. Much as the Justice Holmes quote above sums, "Even a dog knows the difference between being kicked and being stumbled over," so it was with the English common law. If you accidentally killed someone, it is very different than if you intended to do so. Moreover, if you intend to kill in order to further another malicious goal, then the social violation becomes more severe. The same is true between accidentally taking something you thought was yours versus taking something you know is not. For most interactions, these "rules of thumb" are sufficient to sort out who is right or wrong in a given situation. Whether among the native tribes on the American plains or in a medieval English village, if you worked to create something like a coat, or a pair of shoes, and then someone comes along and steals it, within both of these diverse social groups there is a clear recognition that is an unfair action.

But what happens if two people worked to create a pair of shoes and both feel entitled to ownership? A number of outcomes can be envisioned. They may agree to make a second pair, or they may argue and squabble. Indeed, they may even come to blows to settle the matter. As the latter can become socially disruptive, oftentimes third-parties will come in and attempt to mediate the dispute. The trouble there is trying to decide what is fair. Simply leaving the decision up to one person leaves the decision vulnerable to accusations of favoritism and more conflict. The way around this is to create rules for deciding how to mediate a dispute. The rules created are only limited by human imagination. They may be very practical, like calculating the amount of time each contributed to the endeavor. Or they may be exotic and require completing a trial, engaging in physical competition, running a race, or even games of chance like "rock, paper, scissors."

Regardless of method, the creation of a ruleset provides arbiters with an agreed upon set of conditions to resolve a dispute. Likewise, because the ruleset is a consistent source of dispute resolution mechanisms, then the ruleset itself becomes a source of authority to mediate disputes. For a simple example, let us say someone in our little tribe takes a piece of fruit from you. The rule is, if someone takes a piece of fruit, we will do "rock, paper, scissors" to determine who gets to keep it. The trouble with that is what happens if the taker loses "rock, paper, scissors" but still refuses to give the fruit back? In other words, how is the rule enforced? This is where the honor systems mentioned earlier start to come into play. But what if the honor system does not work? One way is to create another rule that says certain members of the group are allowed to enforce rules. Meaning, not only does the initial rule against theft provide neutral authority for what should happen, but another rule can be made to give authority to someone (or many people) to ensure the first rule is followed.

Generally speaking, in small social groups, this task will fall upon an elder, a leader, or a group of leaders to ensure rules are followed. However, the ways rules are enforced are as myriad as the stars. Many Native American tribes practiced a form of restorative justice to reintegrate a member that violated social norms. In those Native traditions, it was the community that was the source of authority.[59] Meaning, the authority to resolve the dispute came through mutual agreement of everyone in the tribe. By contrast, the modern-day Hazda[60] are a largely egalitarian hunter-gatherer society with no hierarchies whatsoever. With the Hazda, disputes that become serious enough are often solved via voluntary relocation of the offender or the offended. As noted above, the medieval English tended to treat violations of social norms through retributive punishment. In the English tradition it was the King, or a designee of the King, that carried out that punishment. Meaning, the authority to enforce dispute resolutions was granted by royal decree. While this may seem a minor point, it actuality, it is a major divergence for sources of rule enforcing authority. This traces back to the earlier discussion around power and control in competitive resource capture social structures.

The notion of land ownership was nearly impossible for early Native American tribes to comprehend. For a person raised in the West, the easiest analogue would be if your neighbor came along and claimed they "owned" the air in your home and told you that you were no longer allowed to breathe it without their permission. This would be an obvious absurdity for a number of practical reasons. For the early Native American tribes,[61] the idea of "owning" land was just as absurd. But in the competitive resource capturing Euro-Christian tradition, owning land was a divinely mandated right[62] and necessity, especially regarding lands occupied by non-Christians. As discussed earlier, land capture, and control in general, was accomplished by force in the pursuit of extortion and plunder. But as forceful control of those territories became codified and boundaried, sovereign authority coalesced around one very simple premise: Only the sovereign is allowed to use compulsory force.

The rough idea being, since the land is owned by the sovereign, the only way to enforce that ownership is by having the ability to force others off the land, or to exclude others from it. Otherwise, sovereign control of territory would be meaningless. Thus, as the idea goes, the sovereign—and only the sovereign—can compel a person to do something against that person's will while they are on sovereign soil. And since it is only the sovereign that wields this power, then the sole authority to use compulsory force rests with the sovereign and the sovereign alone. Of course, the sovereign can, and indeed, must be able to delegate that authority to others for this authority to work. The logical endpoint of this authority to compel expulsion or exclusion from land the sovereign controls also extends to actions taken while on sovereign land as well. Meaning, if you find yourself on sovereign land, of a necessity to the rules, you are able to do so only by the blessing of the sovereign. And while you are there, you must do as the sovereign wishes.

Put together, we arrive at the source of sovereign authority to enforce behavioral edicts through compulsory force. Coupled with retributive punishment for norm violations, we can start to trace out how the adversarial legal system[63] evolved in common law jurisdictions. As a minor side-note, it is worth pointing out here that there are demonstrable degrees of hierarchical social systems on display in the examples provided thus far. The strictly agricultural and animal husbandry based social structures of the English feature the most hierarchical rule sets and rule enforcement. The semi-hunter-gatherer, semi-agricultural Native Americans less so. The pure hunter-gatherer Hazda the least. While this is clearly not dispositive of why these hierarchies exist the way they do in each respective society, it certainly lends credence to the broad thesis outlined thus far. That thesis being, the more reliant upon agriculture and animal husbandry a society is, the more likely it will feature rigid, top-down control structures. The less reliant upon agriculture, the more egalitarian and cooperative the ruleset, and the means to ensure rule adherence. Put simply—and only broadly speaking—competition for survival resources seems to either require, or inevitably ends up with, complex rules and top-down authority to ensure social cohesion. It also appears that cooperation for survival resources generally does not.

CHAPTER FIVE: FORCE

It must be said up front, and especially among pack animals, forcing a fellow member of your pack to do something against their will is universally met with negative feedback. Imposing your will on another person, without their consent, is a fundamental violation of personal integrity. If you attempt to capture a wild animal, it will resist. If you attempt to push someone out of your way, they will resist. If you attempt to force someone to give you something they created or collected, they will resist. If you attempt to restrain someone against their will, they will resist. Even a slight application of force, such as an unwanted, but otherwise benign touch, may provoke a strong negative reaction. Yet, on some level, we all collectively submit to the imposition of sovereign will upon our lives, productivity, the fruits of our labor, and even our offspring. Seems odd, does it not?

The modern concept of submitting to sovereign will is so ingrained in Western society, the vast majority can scarcely conceive of resisting. In fact, and rather inversely, we often find ourselves raging at those who do choose to resist sovereign will. Rioters come to mind. As do thieves, robbers, and other ne'er-do-wells. Much like the English common law, we all recognize that those expressions of sovereign disobedience are unfair to the rest of the populace. Naturally then, those that choose to do so are predictably met with scorn. This brings us back to the earlier discussions. As noted in the previous chapters, once power has been attained in a competitive environment, the next step is to lay down some rules. But in order for those rules to have weight, they will ideally have a semblance of fairness and neutrality. The source of that fairness or neutrality is what provides authority in their enforcement. In common law jurisdictions, this authority was historically broken down into two components:

1. Universal laws; and
2. Prohibitive laws.

Universal laws formed the foundation for the common law. *Malum in se*[64] is the Latin form of the term. Roughly translated it means, "evil in itself." By contrast are laws that are deemed *malum prohibitum*.[65] Roughly translated, that means "prohibited evil." Keeping in mind, of course, when these terms were concocted, the Church still wielded rule-making power and authority in England. Under church and royal doctrine, these laws were deemed universal or prohibitive based on the perceived morality of the proscribed act. The idea being, it is immoral, for instance, to kill someone if they angered you. Likewise, it is immoral to steal from someone. If you do so, then in the eyes of the Lord and the King, you have committed an immoral act. A sin. Modernly, these acts are distinguished by the mental state of the offender. If you intended to kill someone because they angered you, that act is *malum in se*.

Prohibitive rules are just that. They may be enacted for any number of reasons, and most commonly, they are rules enacted for public safety, or for the maintenance of peace. Modernly, jaywalking is commonly cited as a *malum prohibitum* act. But just about anything that is not inherently "evil" can be, and often is, proscribed. For instance, it is illegal to grease a pig with the intent to capture it for fun[66] in Minnesota. In Wisconsin, it is illegal to label butter for sale if it is not "pleasing."[67] Likewise, it is still illegal to wear armor[68] while visiting Parliament in the United Kingdom to this day, despite the fact the law was passed in 1313. Clearly, none of these actions would run afoul of the Church or morality. While these laws may seem dated, or a bit silly, it is worth mentioning that, if you resist a legally authorized agent of the state attempting to enforce any of these rules, all of them are ultimately punishable by death. Technically, you would be getting killed for resisting state authority, assuming you continued to escalate your resistance. But traced to its legal roots, all sovereign laws are backstopped by lethal state violence. Meaning, when it comes to the concept of sovereign power, control, and authority, it is the sovereign and the sovereign alone that is allowed to set the rules of the land. The way the sovereign ensures sovereign rules are adhered to, without question, is through the sovereign authority to wield compulsory force, up to and including killing you. Recall, murder with the intent to murder is *malum in se*. The act is evil in itself. But when contemplated or carried out by the sovereign it is perfectly fine, so long as the sovereign commits that murder under sovereign authority. Technically, the sovereign cannot just willy-nilly murder you. But since the sovereign is in power, is in control, and is the authority, then you really have little say in the matter when the sovereign uses force against you.

As noted at the outset, power is an abstract concept. And much like power, force can also be abstracted. The simplest example is the difference between getting punched in the face and having a big, strong, mean looking person threaten to punch you in the face. Getting punched in the face is a very tactile experience and is unmistakable when it occurs. The threat of getting punched, however, invokes and relies upon the potential face-punch recipient's imagination. Oftentimes, oddly enough, the threat of getting punched[69] can actually be more effective at imposing one's will than an actual punch. This is an important point when we describe sovereign power as force. Whether discussing a bully in a tavern, or a state actor doing a bit of saber rattling on the international stage, the application and projection of actual, physical force is not the only means available to encourage or retard human behavior and decision making. And it is principally the wielding and use of abstract force—the threat of force—that underpins sovereign rule enforcement authority. Boiled to simplicity, the entirety of the English and American legal systems are predicated on the premise that you will do as the sovereign commands or the sovereign will kill you.

That the notion is obscured by institutions such as the courts, or pieces of paper, like the Bill of Rights, does not remove or diminish the core premise. It also serves to explain why the English and American civil and criminal legal systems are based on an adversarial model. While it is quite common for people to rage against the complexity of the legal system and the fine detail, this is a natural evolution of resolving disputes in a competitive environment. In fact, it is the same reason the 2023 American National Football League (NFL) rulebook is 232 pages long.[70] Recall, the core premise of the game is keep and hold a ball while crossing a line while your opposition tries to take the ball away. Meaning, it requires 232 detailed pages of rules to try and ensure that running with a ball is fair. And the overwhelming likelihood is that ruleset will continue to expand in length and complexity. Just like the English and American legal systems have for the last 1000 or so years.

As noted above, the sovereign wields lethal compulsory force. Everyone subject to lethal compulsory force thus has an incentive to try and ensure that is utilized as fairly as possible. And that is where rule exceptions start to arise. For example if, as the English did, there is a rule that says, "If you take something that does not belong to you, the King will cut off your hand." Everyone agrees right up until the King's favorite cousin mistakenly picks up a pair of gloves that looked just like his and walks off with them. According to the rule, the King's favorite cousin should lose his hand. Since the King does not want people to think he is playing favorites and does not want to cut off his favorite cousin's hand, he makes an exception to the law. Now the rule says, "If you *intentionally* take something that does not belong to you, the King will cut off your hand." Of course, the person who lost their gloves can then say, "Hey wait a minute. He *intended* to pick those gloves up and they were not his. You should still cut off his hand." So, our wise King must make another exception. Now the rule says, "If you *know* something is not yours and you *intentionally* take it, then the King will cut off your hand." But then let us imagine the glove's owner says, "He knew they were not his. His gloves are blue. Mine are black." To which the King's cousin says, "No, that is not true. My gloves are black also." Now what is the King to do?

If you are not picking up on what this little hypothetical is getting at, the trouble with competition based rules is complexity. The more that is at stake in *any* competition, the more complex the rules will become. It is inevitable. The lack of stakes is why a group of friends can play football at a Sunday barbecue without 232 pages of NFL rules to guide the game. Of course, anyone who has played a "friendly" game of football with their buddies has probably seen a few squabbles about the rules anyway. But, for the most part, people will recognize the game is for fun and is meant to be a largely cooperative competition done in good spirit. However, if those same guys form an amateur league, the rules become more complex and more important. Likewise, the conflicts become more intense. Under that intensity, rulesets get tested, which leads to refinements. If you put advertising dollars, multi-million dollar salaries and stadium ticket sales on the line, you get 232 pages of rules that people fight over constantly. Just imagine what the NFL rules would look like if there was a death penalty in football.

Thus, a major problem with competitive systems featuring punitive or coercive rules enforced with lethal violence is the enormous complexity required to ensure those rules are fair. The deleterious effect of that complexity is it becomes a source of chaos. Each exception creates another avenue to exploit. Each exploitation creates the need for another exception in a never ending, bottomless fractal.[71] As an example, there was a debate recently about the number of pages in the United States Tax Code. Federal tax statutes require over 2,600 pages of text just for the "official" rules— the actual statutory laws. On top of that, there are another 9,000 pages of Internal Revenue Service (IRS) administrative regulations. But that is not all. To be sure you are in compliance with Uncle Sam, you must also familiarize yourself with another 70,000 pages of tax caselaw.[72] Recall the earlier discussion about authority, one of the key components of common law legal systems is the concept of *stare decisis*.[73] Translated directly, it means "decided." Traced further out, it relies upon the concept of binding, or mandatory authority[74] to ensure consistency in the application of laws. As you may have gleaned, mandatory means just that. So, those 70,000 pages of tax caselaw are, in and of themselves, rules that also must be followed. Donald Trump rather masterfully pointed this problem out[75] in a debate with Hilary Clinton in 2016. Put simply, for people like Hilary Clinton and Donald Trump, that chaos is beneficial to them. For people like you and me, who cannot afford an army of tax attorneys and accountants, that chaos is a pit.

Tax Complexity Keeps Piling Up

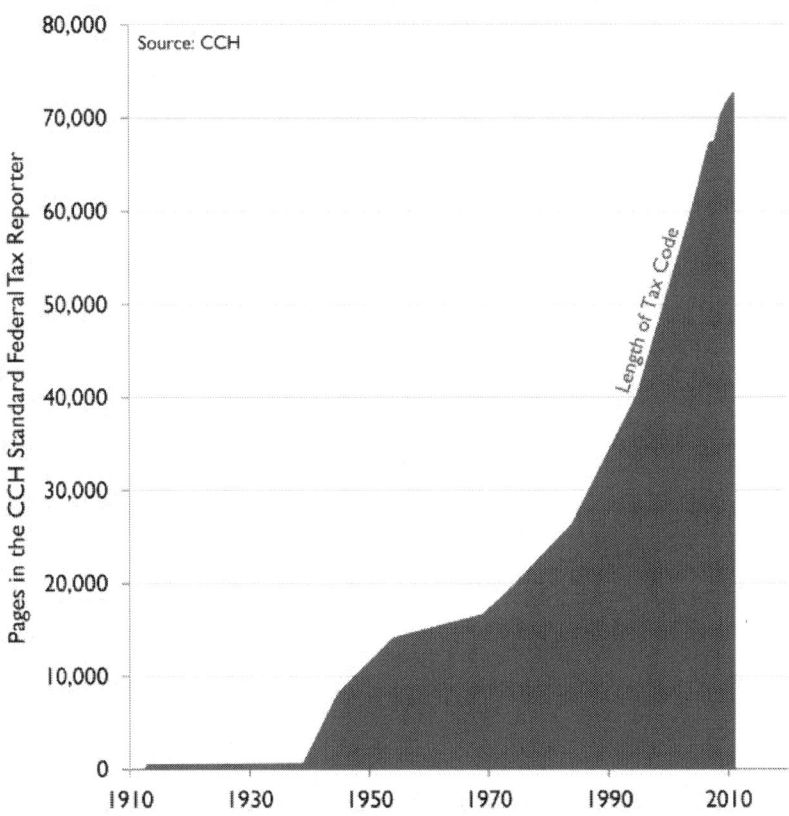

Source: CCH

Pages in the CCH Standard Federal Tax Reporter vs. Length of Tax Code, 1910–2010. Values rise from near 0 in 1910 to approximately 73,000 pages by 2010.

When I said earlier that exceptions create exploitations that require more exceptions in a never ending, bottomless fractal, the image above is what it looks like in practice. Keep in mind, that is just the tax code. The United States Code coupled with administrative financial regulations, securities regulations, banking regulations, commodities regulations, and hundreds of other sources of statutory, administrative, and binding caselaw add up to millions and millions of pages of extraordinarily granular rules. The net effect of all that complexity is chaos. Remember what our good friend Littlefinger from *Game of Thrones*[76] pointed out in the last chapter: "Chaos is a ladder." Which takes us back to the beginning of this whole discussion on power. Recall, the dictionary definitions of power are:

1. Political or national strength: *The Second World War changed the balance of power in Europe.*

2. The possession of control or command over people: *Words have tremendous power over our minds.*
3. Political ascendancy or control in the government of a country, state, etc.: *They attained power by overthrowing the legal government.*
4. Legal ability, capacity, or authority: *The legislative powers vested in Congress.*
5. Delegated authority; authority granted to a person or persons in a particular office or capacity: *A delegate with power to mediate disputes.*
6. A document or written statement conferring legal authority.
7. A person or thing that possesses or exercises authority or influence.
8. A state or nation having international authority or influence: *The great powers held an international conference.*
9. A military or naval force: *The Spanish Armada was a mighty power.*

From those dictionary definitions, I pointed out you can extrapolate some common themes about power:

1. Strength
2. Control
3. Authority; and
4. Force.

As we wrap up this chapter and this section of the book, let us see if we can boil the concept of power down to some semblance of coherence. The modern monetary system—and the global political economy that supports it—is based on a set of rules that have developed over several hundred years. Those rules are in place because the system that prevailed is based upon violent competition for concentrated, human created resources. The unchecked violent competition for those resources created unsustainable chaos. Groups that successfully banded together were able to demonstrate their strength by bringing order to that chaos. From that expression of strength, they gained control of those concentrated, human created resources. Once in control, they created rules to govern and compel the continued creation and distribution of those resources. The rules they adopted were to manage a competitive, tournament style, fee-based system that is weighted by the relative strength of the players within the power hierarchy.

However, because this structure was rooted in and based on a high-stakes competition, unethical behavior, rule testing, and rule violations required increasingly complex rules to maintain fairness among each new evolution of players. That complexity led to internal chaos, which provided opportunities for further exploitation of the rules by new or emerging players. From that chaos, strength then manifested within the system through the capturing of sufficient resources to exploit the chaos of that systemic complexity. This naturally led to more complex rules that, over the last few hundred years of refinement, have resulted in the current system we find ourselves in. Meaning, for all intents and purposes, the game is being played exactly as it was laid out.

Or, put another way, the modern global monetary system and the political economy that supports it is working from deeply flawed logic. That flawed logic being twofold: the notion that violent competition for resources is natural and that rules can be made to prevent cheating in that competition. The trouble being, the infinite ways cheaters can work to circumvent well-intentioned rules, especially when they are incentivized to do so by the structure of the game. This is much like a computer program that gets exploited.[77] With any computer program, it is necessary for a human to create a process that can be logically executed. The trouble arises when the programmer's logic does not fully correspond with their intention. They may write a line of code that they think will perform a certain function in a certain way. In doing so, however, they may introduce an undetected way to execute a process that goes against what they intended. Meaning, they fail to understand all the implications of the logic they are applying. What the exploiter does is find that hole in the programmer's logic and uses it to their advantage against the intentions of the programmer. For example, the Nomad crypto bridge[78] exploit was enabled by the programmers because they incorrectly set a parameter to "0" when it should have been a "1." This is what it looked like parsed out.

> *The vulnerability appears in a scenario when fraudulent messages, not present in the trusted **messages[]** map, are sent directly to the **process()** method. In this scenario*

messages[_messageHash] returns a default null value for non-existent entries so the **acceptableRoot()** method is called as follows:

```
require(acceptableRoot(0), "!proven");
```

In turn, the **acceptableRoot()** method will perform a lookup against **confirmAt[]** map with a null value as follows:

```
uint256 _time = confirmAt[0];
if (_time == 0) {
    return false;
}
return block.timestamp >= _time;
```

*As we mentioned in the beginning of this section, **confirmAt[]** map has a null entry defined resulting in **acceptableRoot()** returning **True** and authorizing fraudulent messages.*

This is essentially what happened in the example with the King's cousin above, just in computer language rather than human language. So long as everyone did what they were supposed to do—based on the programmer's intention—the system worked fine. But, as mentioned previously, if the name of the game is to capture resources, then there is an incentive to do so. If someone leaves high-stakes resources vulnerable by typing a "0" instead of a "1" then there is a clear incentive to try and capture those vulnerable resources. Mind you, the exploiter did not overcome the rules. They simply played by the rules established by the programmer. That the rules ended up with a result the programmer did not want is irrelevant. But this traces back to the earlier discussion about the King's cousin (mistakenly?) picking up a pair of gloves. Recall, the rule was, "If you steal, you get your hand cut off." Keep in mind, of course, this rule works fine—as long as no one picks up anything that does not belong to them. But when that logic proved problematic with the King's cousin, they changed the rules. The new rule said, "If you intend to steal something and then do steal something, you get your hand cut off." The intention there is obvious. It is attempt to close a logical hole in order to prevent theft. The result though, is it increases the thief's incentive to steal. Why? Because the new logic can be exploited by obscuring their intention to steal. If they get away with the theft, they are enriched. If they do not, they can wiggle out of losing a hand because of the flaw in the logic.

Put simply, "Rules are made to be broken."

As Lord Acton once quipped, "Power tends to corrupt and absolute power corrupts absolutely." As I hope you can see more clearly now, there are very practical reasons why this quote resonates. When survival resources are concentrated, those resources become a natural target for competitive control. Unchecked competition for control of resources leads to chaos. Chaos becomes an avenue to demonstrate strength and acquire power. Once in control of those resources, the power structures that form tend to favor top-down administration and rule-making authority. But because the competition to acquire and control resources is antagonistic and the stakes are lethal, the rules to ensure the competition is fair become increasingly complex over time. That ever increasing complexity becomes a new source of chaos. Each new source of chaos then becomes a new avenue to demonstrate strength and acquire power.

The concentration of survival resources in the hands of a few can only lead to competition and conflict. Competition and conflict can only lead to rules. And, unfortunately, history has shown that humans are really bad at making rules. Worse still, really bad rules create more and varied competition and conflict, which leads to even more complexity. And because the complexity is becoming ever greater, those best able to exploit the logic of the game becomes smaller and smaller over time. This predictable outcome, almost unavoidably, ensures that fewer and fewer players can continue to accrue and concentrate survival resources without resorting to increasing corruption and malfeasance. In turn, this ultimately ends in the catastrophic collapse of the game. It is much like golf: A game we can all play and no one can truly win.

At this point, I would also like to call attention to Hanlon's Razor,[79] which says, "Never attribute to malice that which is adequately explained by incompetence or stupidity." I think that is most apropos to the discussion at hand. All of these rule-bending and rule-breaking actions are inherently logical. If you are in a high-stakes, competitive system, it always makes sense to test boundaries. While we may rail against it, none of the actions the power-hungry take are necessarily born of ill-intent or malice. It is completely understandable how the power-hungry throughout history can be perceived as evil. Indeed, many of the actions they have undertaken throughout history were or did, in fact, prove to be evil. The tiny flaw there lies in our perception of what is or is not "evil."[80] The trouble being, if you asked any of those power-hungry figures why they did what they did, chances are good they genuinely believed, or at least convinced themselves, that they had a solid, ethically-based reason for their actions. And as we will see in the later section on Greed, many great figures in history may well have started off with a noble purpose, but can very easily and predictably end up monsters. Nevertheless, now that we have shaken up the conversation around power a bit, let us move on to the mother of greed, Money.

PART TWO: MONEY

CHAPTER SIX

Long before we can begin to coherently discuss Bitcoin, we must first understand where money—*the idea of money*—comes from. This chapter must, of a necessity, be condensed. The topic of money is deep and interwoven. Plenty of ink has been spilled roughing out what money is and where it came from. None of it is fundamentally correct. No one has comprehensively or definitively defined money. There are many ideas about what money is, how it works and, indeed, why it exists at all. At the end of each line of inquiry is the unsettling realization that money is simply a construct. It morphs and changes with circumstance. There is nothing natural about money. There is no innate quality that money possesses. Yet we carry on with the illusion that money has fixed properties and can be understood as such.

It is most peculiar.

This is partly why much of what follows will likely prove counter-intuitive to many. As the late David Graeber pointed out, concepts of money are so ubiquitous[81] today that few question the origin story. Indeed, most cannot even imagine a monetary system different from what exists today. The trouble is many of the assumptions around money, money creation, and money's role in the world are deeply flawed or are unsupported suppositions. Those flaws are largely a symptom of the momentum and reach of sound-bite reasoning and faulty heuristics. Economists have done their share to further muddy the waters. The Austrian School[82] comes closest to an understanding of modern money. Even there it is not perfect. The problem with economics in general is the entire school of thought revolves around very generous assumptions. These assumptions are rarely challenged. They are even more difficult to unpack. If you ask someone, "What is money?" you will likely hear any number of definitions. Medium of exchange, store of value and unit of account are the Economics 101 standards. Generally speaking, these definitions come from core concepts rooted in Adam Smith's 1776 book, *The Wealth of Nations*.[83] For those unfamiliar, Adam Smith's money "origin story" goes something like this:

> *"Once upon a time, people would trade things like shoes or oxcarts. As life and things became more complex, it became too hard to trade these things. It was hard because there is no easy way to divide an oxcart or a shoe. To fix this problem, these people created money. Then from money came lending. This naturally turned into credit and debt. Rinse and repeat for a few hundred years and we get the financial world we live in today."*

This idea holds a certain logical appeal and, on its face, seems perfectly sound. There is the minor inconvenience that Adam Smith simply made it all up. If you run this story by an anthropologist[84] you will discover it is a bunch of malarkey. They have found no evidence of a society on earth developing this way. Not to say that has not happened. It is to say it seems rather unlikely. Certainly barter economies have existed in the past and even in the present. But those barter systems usually crop up after the collapse of whatever previous monetary system was in place. It is also true that people have used things like coins, shells, and beads to exchange value for time immemorial. But to assume that one naturally flowed from the other is a mistake. I rather argue here that the way we think about and use money today is relatively recent.

Rather than a system of barter that organically grew into fungible[85] money as Adam Smith said, I posit what we actually have is a system designed to extract taxes[86] from a population. To better understand this position, I hope you will indulge a little tweak to the broadly accepted money paradigm. To accomplish this, let us first discuss what happens when people exchange value, whether it be time, effort, or things. In furtherance of moving this process along, let us engage in a little thought experiment. Imagine a close friend invites you to their house for dinner. You arrive at the agreed upon time. Your friend has prepared a meal that you share. Would you expect your friend to present a bill to you at the end of the meal? I am sure it has happened at some point, but I think that would be unusual. Now, if two weeks later, your friend fell on hard times and needed some food, would you charge them? I would certainly hope not.

Likewise, imagine when someone asks, "Pass the salt please." Is your first response, "My fee for passing the salt is $0.50?" Probably not, right? As a general—and I use the term lightly—first-principle then, let us say at a minimum that human beings exchange value. Whether we call it sharing, selling, trading or something else is immaterial. There is conceptual research[87] showing cooperation is an essential human trait. That trait has contributed greatly to our success as a species. A key part of cooperation involves the exchange of utility, services, or things of perceived value. With that in mind, I think we could also reasonably assert that, among close associates, most trade is frictionless.[88] We exchange freely with our trusted associates, friends and loved ones. The inherent expectation in these exchanges is fairness. The unspoken trust is that our sharing and trading will be roughly equal measures of give and take. In fact, there is a growing body of research[89] demonstrating people orient towards three basic patterns: Givers, matchers, and takers.

For a case on point, we all have had friends that take more than they give. Assuming we bother to keep them as friends, if they come asking to borrow something we value highly, we may require collateral to ensure we get whatever that is back. That collateral may be as gentle as social pressure. It could also just as easily be a signed contract or an escrow agreement. Likewise, we probably all have friends that give much more than they take. These friends may range from the doormats to the grandiosely generous. If they ask to borrow, we may simply give freely out of pity or in recognition of their previous generosity. It is scarcely contentious to say our close personal relationships are often mediated by these interactions. How we treat an exchange varies based on our trust in the person we are dealing with. If I know you are a fair, reliable close friend, I will exchange value with you one way. If you are a friend that is always borrowing and never reciprocating, the exchange will be different. The point is, the further away from that "trust circle" we get, the more insurance we may demand. This is one way that we can ensure the value given will be matched or returned in kind. The less we trust someone, the more insurance we may require, or the more of a doormat we may become.

Early Monetary Systems

For most of human history, human social groups tended to max out at around 150–300 people[90] in a given "sphere of influence." Generally speaking, value exchange in these small communities was done by informal contract. If you need shoes and I make shoes, I simply give you a pair of shoes. In this is the implicit (and sometimes explicit) understanding that you now "owe" me an equivalent value for a pair of shoes. The Austrian School of Economics[91] gets the value component of this transaction correct. Unlike modern economists, the Austrians recognize that value cannot be "intrinsic." Instead, value is inherently and irrecoverably receiver dependent. There is no "supply and demand" in trade. It is just "demand and demand," with each side "demanding" of the other. Even this may be a bit of a misnomer. Hoping or desiring could just as easily supplant the term. All the same, this aspect of perspective-dependent value allows for exchanges such as the hypothetical shoe trade above to work. This is very different than the Adam Smith version. His version roughly says value must, at all times, be bartered equally. In the actual, and dare I say observed, implementation of trade, value is quite malleable. Where the economists' models presume "rational actors" exchanging value, behavioral economics has long since discounted any such notions.[92] For example, a pair of shoes might be extraordinarily valuable to the receiving party. In exchange, they may return anything from a hug to an oxcart, or even $20 million[93] cash. Moreover, anything between a hug and $20 million may prove sufficient to the receiver. If both parties view the transaction as fair, it will result in both parties happily exchanging value.

Of course, in an ancient Northern European village, such an arrangement is ripe for the problem of free riding.[94] Free riding is the large, community-scale version of the friend who takes and never gives. One way free riding was curtailed, particularly in many early medieval European villages, was the use of a ledger.[95] The value tracked in the ledger could be anything from help with a barn raising to a dozen eggs. Each family would keep track of goods and services exchanged using their own ledger. In many cases, an exchange would not even require an entry in the ledger. This is because most "debts" would be immediately canceled during the exchange, or shortly thereafter, e.g., "if you help me catch my pig, I will give you a beer." But if a particularly valuable thing or service is being exchanged, the parties may write the details down in a contract. The interesting thing about a contract is, the written promise itself—again within a trusted circle—can also become a valuable commodity of exchange. A promissory note,[96] if you will. In this way these contracts would also become exchangeable and tradable beyond the two people making the initial agreement.

If I thought you were trustworthy and I gave you a side of beef, but you did not have anything I need, we could simply write those details down. The writing could be as simple as, "You owe me for a side of beef." I could then use that written instrument to trade your debt to me with someone else in the community. This way I could get something I need, like a farm tool, or help with repairing a roof from the same side of beef I gave to you. The neat thing is, the tool maker or the roofer might make the trade even if they did not need anything from you either. Your good reputation and your written promise to provide a "side of beef value" would allow further trade to carry on. As alluded to above, this is the crude and original form of "promissory note." It looks something like this in practice:

1. Adam owes Baker for something of value and writes down what the debt is for.
2. Baker takes that promise and offers to trade it to Charlie for something of value that Charlie has and Baker wants.
3. Provided Charlie knows Adam is good for the debt, Charlie accepts.
4. Charlie could then choose to "redeem" from Adam, or he could continue the chain by trading Adam's promise to Debbie for something Debbie has and Charlie wants.

Round and round these promissory notes would go until at some point the community gets together and has a reconciliation. During the reconciliation period, the ledger accounts of who owes what to whom are balanced. Any value given and not received is made whole, or at least as close as possible. Also during this time, if a perceived or real mismatch in expected value is discovered, a select group—usually trusted elders—would mediate the dispute. Through this, the community would ensure relative fairness based on local standards.

Once the books were settled, any trivial or difficult to reconcile balances could be made whole using any divisible device, but generally this was done with specie[97] (coins, usually gold or silver). Historically speaking,[98] that would have been the primary time specie would be used to exchange value within these small communities. There are even modern examples of the "books" being reconciled with things as diverse as cigarettes, ramen noodles and pecan pralines.[99] Remembering, of course, this all depends on the existence of high-interpersonal knowledge and trusted relationships in the community. In small communities to this day, from small towns to prisons, it is not uncommon for people to know each other well and trade accordingly. "Just pay me when you have it" is a refrain that is quite common in small communities. Indeed, such an instance recently occurred for me personally when I discovered the muffler shop that serviced my car did not take credit cards. Nevertheless, what this all demonstrates within high-knowledge, high-interpersonal trust trading systems, is that they are not always quite direct barter. They are also not quite trading with dedicated mediums of exchange. It is often a little bit of both, with some form of ledger in-between. At the core, however, all of this exchange is, or can be, facilitated by some level of trust and presumed fairness based on reputation and inter-personal knowledge.

The little hiccup here for the medieval villagers is trade would, of a necessity, be limited to that small group of people. The value of a promise carries less weight the further away you go. If the promisor is not close enough to know and trust well, let alone collect from, this type of informal ledger trade can rapidly break down. The same problem would present itself if a stranger came into the community. A far-flung traveler passing through town would have no ties to the community. Absent direct settlement, e.g., barter, a trusted trade with them would be highly risky. The rational course of action would be to assume the stranger would likely be a free-rider right out of the gate. The solution to this problem, whether in a prison, or a medieval village, is also the use of divisible "currency." In the case of the medieval villager, rather than demand direct barter from a passing stranger (like labor for food), they could instead just take a few bits of specie and call it good. In turn, those coins would be helpful the next time the group needed to balance their accounts. Likewise, when a member of the local group traveled afar, they too could use specie at their destination to procure goods from another community when they themselves were the stranger. Of course, all this begs the question, "if this worked so well, why did everyone suddenly start using specie or other divisible currency for *all* trade, even among trusted friends?" To answer the question, we have to better understand the real problems money is trying to solve from the perspective of the people that create the money.

Imagine you are an early medieval landlord and your King tells you to tax your peasant population and send him a cut. The easy way to do this is to simply go around and take a certain amount of whatever they produce—by force, if necessary. Indeed, if traced back far enough,[100] the "great" royal dynasties of Europe are essentially rooted[101] in crude protection rackets that grew out of the prehistoric "agricultural revolution." While difficult to accurately trace, the Hallstatt culture[102] appears to be the genesis. The behavior of the Hallstatt ancestors to these royal dynasties was likely not dissimilar to what gangs and street thugs do today.[103] The concept is simple: If you are good at violence, it is far easier to extort and rob people than it is to make or grow stuff. If you are really good at violence, you can often get other people to do the violence with, or for you. Kings, as it happens, became kings because they were the best at organizing and coordinating that violence. The better at organized, coordinated violence they got, the bigger their kingdoms became. This was especially true if they were both excellent at violence and still maintained an air of "fairness" to their extortion and robbery. Even better still if they are willing and able to step in and mediate disputes.

Like a friendly neighborhood mobster, if you will.

At any rate, after the King's landlord collected those taxes from their peasants, the landlord would keep a share for himself. The rest would go to the King in order to settle the landlords' tax[104] to the sovereign. In this way, community produced resources were redistributed up from the working peasant-class to the managerial landlord-class and then on to the sovereign. If you did not notice, this is much the same redistribution process in use today. With the exception today, of course, that it is much more layered and theoretically voluntary. It is important to the discussion at hand to keep in mind that there is still a gun at the end of that chain. Nonetheless, the larger trouble with taking "stuff" is that there is only so much wheat, or meat, or shoes, or what-have-you that you can store before it rots, loses value, or gets too cumbersome. Meaning, direct taxation by theft or extortion of goods can quickly become a basic logistics and storage problem. Not to mention, the direct taking of goods leaves a lot of value exchange between your peasants that you cannot profit from through the theft of "stuff."

Think of it from a King's perspective. If building a fence is a valuable service, why should the King not get a cut of that too? The peasants, after all, are his property by divine right. Their labor is merely a rental, much like a plow horse tilling another man's field. Certainly the King deserves a fee for divinely providing that rental service, right? Nevertheless, the clever way the inventors of modern monetary systems got around these problems was to demand taxes be paid in specie instead of "stuff." Such an arrangement makes a lot of sense if you are a King and your kingdom has grown quite large. Forcing your population to pay taxes in specie instead of "stuff" forces them to trade in specie. Put another way, they are now forced to make and do things in order to get specie to pay their tax to you. This arrangement does three really great things for you—Our most honorable, wise and divine King:

1. You do not have to store a bunch of stuff you do not need right away.
2. You force your population to be measurably productive (no more hiding potatoes).
3. You can give your peasants the specie you took from them in tax receipts to pay them for the work you are indirectly forcing them to do, or to pay for the products you are indirectly forcing them to make.
4. Which, in turn, incentivizes them to be even more productive

Put simply, if you make shoes, then you must sell some of those shoes for specie, or you cannot pay your tax. Labor must be compensated in specie, or the laborer cannot pay their tax. Landlords must collect specie from their peasants, or they cannot pay their tax. In turn, the sovereign can now buy things the sovereign needs or wants using the same specie they collected from their peasants in taxes. Since the taxes are only a percentage of what the peasant makes, it encourages them to increase how much they make in order to make and save specie. It is a pretty neat trick.

Going larger still, all sovereigns caught onto this little scheme around the same time. The printing press went a long way in furtherance of this concept, as the thinkers of the time like Adam Smith, were able to better and more widely distribute their thoughts on these matters. In any event, trade in specie was largely adopted because gold and silver already enjoyed a widely recognized "value" among people and sovereigns alike. Once trade in specie was formalized, it also enabled the sovereigns to trade with each other in a coherent manner. Just like that, you have created a relatively low-friction, hierarchical market economy that survives in this general form to this day. In short, and to sum the answer to the question, "what are the problems money is trying to solve?" I would answer that, for the population and the sovereign respectively, money:

1. Facilitates the exchange of value in low, or zero-trust situations; and/or
2. Allows for the bulk extraction of fractional value from the personal or collective industry of sovereign subjects through taxation.

This definition, of course, is quite different from what Adam Smith or, indeed what most economists would say. Yet, from a historical perspective, it is a more accurate one, especially in the industrialized West.

CHAPTER SEVEN: COMMODITY MONEY

As noted above, through most of medieval European history, trade among trusted groups tended to occur through formal or informal contracts and ledgers. Generally, specie would only be used for a collective final settlement much later on. The rough path being, I need or request something from you, you supply it to me if you want. Depending on the particulars, some form of contract arises between us that says I will repay you in like value in the future. If the transaction is large, or complex, we write it down on a ledger. Everyone maintains a ledger of accounts. At some point, we all balance our ledgers and start all over again. With that brief overview in mind, the evolution of money is obviously more complex than what has been outlined thus far.

What I aim to capture here are the broad historical strokes as they apply to the industrialized, pseudo-capitalist economies of today. In other words, the financial system that was born from the Euro-Christian nation states that arose from the _Treaty of Westphalia_[105] during the late Middle Ages. As noted at the outset, there is no unified system, or theory of money. Formalized trade systems have arisen all over the world throughout history. The way they were created and functioned is far more attributable to where and how they lived, where and how they get their food, and the Gods[106] they worshipped as a result. This is as true of the developed western economies today, as it was true of the economies of the American Plains Indians, the ancient Mayans, or the Eastern dynasties. For hopefully obvious reasons, the focus here is on the western economies in which we currently live.

With that lengthy caveat in mind, it is around the time of the Renaissance that we start to see specie being utilized in a very similar fashion as paper money today. Coins of various denominations circulating widely and nearly all formal trade being settled via the medium of standardized specie. Much like today, if one were particularly industrious, they could accumulate coins for a large purchase. Or, those savings could preserve their purchasing power[107] to cover a rainy day. With a system such as this, it also becomes possible to balance a ledger by simply counting the number of coins one owes to another, or vice versa. This, of course, is a much simpler method of accounting. That becomes especially true in an increasingly complex system. Beyond a small village, the need to account for the particulars of each transaction would become prohibitive. Likewise, the period of reconciliation for personal promissory notes would be untenable at larger scales without complex forms of automation and error correction.

As you may have guessed, this brings us to a point where fungible money becomes a medium of exchange, a store of value, and a unit of account. This is, of course, the same destination where Adam Smith arrived. Our journey, however, wove a very different path than Mr. Smith's money origin story. It is an important distinction, because Adam Smith's version suggests the origin of money was benign and simply born out of natural utility considerations. The perhaps more accurate version for the industrialized West instead demonstrates that sovereign money was anything but benign. It was rather a deliberate tool to more efficiently facilitate theft and extortion from the start. Of course none of this explains *why* gold or silver might be valued as a unit of voluntary exchange. With the sovereigns in particular, this is a question one should certainly ask. After all, it is not at all clear why gold or silver should be valued above any other fungible item. Author Lyn Alden[108] proffers an answer in her book *Broken Money*. To wit, she writes:

> *Why did gold and silver defeat all other commodity monies to reach the modern era as usable money? The answer is that*

these were the two that could maintain high enough stock-to-flow ratios against the rise of human technology.

Alden describes stock-to-flow as an important concept when it comes to understanding money. The core idea is the stock is the supply of the commodity—the amount available to use. The flow is how much new supply is created. At face value this is all perfectly logical and reasonable. Gold has a limited stock that has been mined and refined. On the flip side, new gold issuance from mining increases this supply by somewhere between 2–5% per year.[109] Moreover, gold cannot (yet) be "counterfeited" and can be assayed to verify the material presented is actually gold. Of gold, Alden assumes a lower supply rate (1.5%) and arrives at a stock-to-flow ratio of 67:1, describing it as "the highest stock-to-flow ratio of any commodity." Alden then outlines a number of other commodity "monies" such as shells, beads, tobacco, feathers and the rai stones of Yap in furtherance of the stock-to-flow theory. The central tenet of this theory says, in effect, that because technologically superior Europeans were better able to produce these monies than the aboriginals (making them high flow), gold and silver eventually pushed those monies out of favor. Framed more disparagingly, the Europeans could counterfeit Native money with ease, so Gresham's law[110] took effect and pushed out the "bad" money (e.g., shells) in favor of the good money (e.g., gold and silver.)

I deeply admire Lyn Alden. I think she is a thoughtful and engaging economic historian, an excellent writer and speaker, and a highly credible person. I also think this stock-to-flow theory is a bit of poppycock, especially as it relates to non-European "monies" such as shells, beads, and indeed, even the rai stones of Yap. For instance, after first stating that shell money had been used in many parts of the world, Alden goes on to say that some uses for these shells were "more" ceremonial, while others were "literally" transactional money. One particular version of Wampum, a widely distributed cultural item on the Eastern American continent, is then described as "the more ceremonial variety." She notes the colonists incorporated this Wampum into their burgeoning frontier economies. This latter part is absolutely correct. As George S. Snyderman[111] wrote in 1954:

When he arrived among the aborigines of North America, White man brought with him the system of thought current in

> *western Europe. The economy he sought to transplant in the New World* **depended on the existence of highly developed markets and the circulation of a medium of exchange with a fixed or standard value.** *Expansion of trade and establishment of colonies during the seventeenth century seem to have been accompanied by a lack of bullion.* **This, coupled with a now untenable theory that all goods were valued in terms of precious metals, not credits, use, etc., kept the already small supply from fully circulating.**

At that particular crossroads in history, Portugal and Spain controlled the vast majority of the gold and silver supply coming out of the so-called "New World" and jealously hoarded it. This served to further constrain the availability of specie in the colonies. I would also point out the reference to the "untenable theory that all goods were valued in terms of precious metals." Given the historical use of ledgers in the past to facilitate trade, it is clear from this accounting of colonial economics that such notions had, by that time, been entirely lost to the reasoning of Adam Smith and his ilk. As an aside, I would posit those concepts are still lost upon the ruling classes and their subjects today. That said, what happened with Wampum was essentially two-fold. First, the European Christians very quickly realized the Natives really liked Wampum and were willing to trade land, furs and any number of other valuable things to obtain Wampum. Second, market attuned colonists could not acquire adequate sums of specie to facilitate the markets they desired, so they sought a convenient substitute in their New World. But this does not necessarily make Wampum "money" that might be supplanted by the superior stock-to-flow characteristics of gold and silver. Indeed, given the choice, the colonists would have most certainly preferred their common weights and measures to carry on their trade and build their economies. Thus, it rather seems that, for the aboriginals, their cultural artifact just happened to get caught up in the European fungible currency, market-based, resource capture game. As Snyderman writes:

> *The Indians had an object with some of the qualities which White man attributed to money. White man therefore adopted it as money. Since the Indians were so anxious to get it for ceremonial purposes, White man fooled himself into believing*

> *wampum was Indian money. There is no evidence that the Indian ever needed, wanted or used a medium of exchange prior to the arrival of White man.*

Snyderman goes on to say:

> *Wampum was merely a commodity which the Indians needed to maintain all types of social, political and religious relationships and ceremonials. Wampum was an essential to all these.*

The point being, Wampum was not "more" of the ceremonial variety. Wampum was never money to begin with, at least as far as the American aboriginals were concerned. So while it is true that "In all the colonies wampum was used by the Whites among themselves as an interim or emergency form of legal tender," it does not necessarily follow that the superior stock-to-flow characteristic of gold and silver caused the demise of Wampum use by the European Christians. Quite literally no one was settling accounts or trading with Wampum in Europe. They were selling it. Had the colonists had ready access to gold and silver coins, they would have used them. More broadly speaking, what this example highlights is not the supplanting of one commodity money for another due to superior characteristics. It is rather more accurate to say the colonists were using a stop-gap measure to settle domestic accounts that also had some use in capturing resources from the Natives. At the core of the exchange, however, is the rather unflattering prospect that they were outright counterfeiting a cultural item to capture resources from the Natives. That they mistakenly attributed monetary qualities to that cultural item and then chose to rely upon that misattribution to their detriment hardly confers a causal relationship to the stock-to-flow model.

Much the same could be said for the rai stones of Yap. Of these stones, Alden says their high stock-to-flow ratio "is a main reason for why they could be used as money." Oddly enough, just a couple of paragraphs later, Alden quotes an article in *Smithsonian Magazine* that says:

> *The Yapese were not much interested in sweating for the trader's trinkets that were common currency in the Pacific (nor should they have been, a visitor conceded, when "all*

> *food, drink and clothing is readily available, so there is no barter and there is no debt"), but they would work like demons for stone money.*

Much like their northern aboriginal counterparts, it rather appears the Yapese found themselves being similarly exploited by the resource capturing European Christians. As Scott M. Fitzpatrick[112] says, "No one theory or model can fully explain the complex interactions occurring between the various island societies involved with stone money quarrying and Europeans who later became involved in this exchange system." Fitzpatrick goes on to write, "The manufacturing of stone money by Yapese Islanders in western Micronesia is one of the most archaeologically dramatic, but least understood instances of 'portable' artifact exchange in the Pacific." What is clear is that, shortly after the arrival of the Europeans, the islander's "trade" in rai stones accelerated dramatically. Much the same can be said for the fortunes of the Irishman that ferried Yap stones to and fro, himself being one of the Europeans that "were primarily concerned with making national claims to lands for colonial expansion and establishing trading posts to increase their wealth and influence." Again, and much like Wampum, there were no accounts in Europe being settled or priced in Yap stones. That Europeans came across a means to capture resources from distant lands via the use of a cultural item does not necessarily impugn monetary status upon that cultural item. It certainly does not implicate any stock or flow characteristics that could be supplanted by the gold and silver commodity money that all of these Europeans truly desired.

Which ultimately, and rather disappointingly, routes us right back to the beginning of the discussion. Again, and with all respect to Alden, this is not to state unequivocally that stock-to-flow or any other theory is wrong. It is rather to stress the notion that money is simply a construct. Keeping that in mind, if stock-to-flow does not necessarily describe why gold and silver might become a desirable resource for sovereign hoarding, theft and extortion through taxation or otherwise, then what does? The Bible explicitly mentions gold and silver as money. Given the enormous power the Church wielded during the Middle Ages, it would be no surprise to find that gold and silver were highly prized simply from its mention in the Holy Book. The trouble with that is, gold and silver were also highly prized two or three thousand years before the Bible was written. Meaning, it seems much more likely the authors of the Bible simply accepted gold and silver as valuable because it had been that way for as long as anyone knew up to that point. In fact, and as best as can be surmised, precious metals seem to have an almost universal appeal throughout recorded history. Whether in China, India, ancient Rome, the Middle East, Western Europe, or the Americas, people just seem to like gold and silver. That these metals are relatively scarce and relatively difficult to acquire and refine also seems to add to the perceived value equation. With gold in particular, among the metals, it is one of the easier to work with and shape, roughly on par with toxic lead. Gold does not corrode when exposed to the environment. It can also be reused infinitely without degradation.

The gold in your watch or wedding ring today could very well contain gold worn by an Egyptian at the time of the pyramids. Yet a precise reason for why gold and silver became valuable among disparate sets of humans remains elusive. The best answer as to why precious metals—and especially gold—have been used as a valuable, tradable commodity very well may be because, unlike gems, you can:

1. Readily reshape and divide gold without waste.
2. It is stable and non-toxic.
3. It is essentially indestructible; and
4. People think it looks nice when refined.[113]

Much the same is true with silver, save for the fact that it tarnishes. That is quite literally as good of an explanation as I can find. This is especially true given the fact that I put scant weight into the stock-to-flow thesis. The broad point being, from the Chinese dynasties that date back to pre-Roman times, to the Romans themselves, and indeed, an ocean away, the Aztecs and Incas all refined and used gold through the ages. The advent and rise of market economies and value exchange via currency, however, very much appears to be a solidly Western construct. And as noted throughout, much history is skimmed here. From the rise and fall of Rome, to the spread of agriculture to the hunter-gatherer tribes in Northern Europe, to the domination of Christianity in Europe, market money has evolved and changed and shifted with the ages. To reiterate an earlier point, there is no innate quality that money possesses. Likewise, there is no comprehensive definitional understanding of money. Nevertheless, and to return to the topic at hand, in the late Middle Ages in Europe, as societies became more complex, specie became more refined in appearance and weight. To be sure, sovereigns routinely tinkered[114] with those weights and measures. They would change size and purity. They would recast coins at different weights. At other times, they might ban one form in favor of another. Almost universally, these changes were an attempt to deal with spending more than the sovereign collects in taxes and plunder, i.e., deficit spending.[115]

The simplest way to explain this is: If you are a sovereign and you need more gold than you have to buy goods and services *and* you need to keep minting new coins for trade, the easiest way to do both is to take all the coins you minted before and make each one weigh less. Then, as the idea goes, you either trick or force your population to trade in the new coins as if they were the same value before your goldsmiths tinkered with them. Of course, the peasants and merchants often caught on to this little scheme and did some clipping of their own.[116] This is one of the major problems with using precious metals as money. Regardless of who is driving the debasement, one can only recast a coin into a smaller one, or mix it with base metals[117] so many times before the population either demands more of it as payment, or simply stops using it. Is the King of England recasting your coins? Use Italian or French ones instead. If those dastardly Frenchmen start doing the same thing? Well, just switch to Swiss coins, or German ones. This is the essence of Gresham's law, which says, "Bad money will chase out the good."[118] It is also an especially vexing conundrum for commodity money. There is a weight-value to the underlying precious metal in most forms of specie. Once the face value (the denominated value) deviates significantly from the weight-value, or "melt value,"[119] people will quit using it. Instead, the people forced to transact in the debased coins will spend the debased versions first and horde (save) the purer, more valuable ones.

As a side-note, if you have ever wondered where the origins of European hatred towards Jews first manifested, it largely came from monetary debasement[120] and money lending. Keep in mind that, in medieval Europe, the Catholic Church still held a very powerful position of authority. Those familiar with the Bible will know that lending money for interest (usury) is a forbidden practice.¹²¹ The Jews were not constrained by this, at least insofar as dealing with gentiles. Thus, the European kings would rely on Jewish bankers to engage in usury (usually borrowing) and debasement¹²² via clipping, re-alloying and recasting. This gave these sovereigns both a way around ecclesiastical law,¹²³ and a very convenient, non-Christian scapegoat for when their monetary schemes inevitably fell apart. "It was not I, but rather those dastardly Jewish bankers that have ruined your money dear peasants. It is not my fault. Hang them instead of me." Obviously not a direct quote, but that is the gist of how medieval royalty very easily avoided taking responsibility for running up too much debt and debasing their subjects' purchasing power. This is a very simple and obvious benefit for the sovereign. Demonizing Jewish bankers was a no-brainer for Christian Kings and their loyal Christian subjects. They would often do this through mass propaganda campaigns against the Jews. For them, it was a small cost to ensure His Imperial Majesty got to keep his majestic head affixed to his royal body. Meanwhile, he would still get to fund his next whim, whether it be a war, or a stunning new home.¹²⁴

Returning to the topic at hand, the sovereigns of these nation-states could have easily banded together and coordinated their monetary tinkering. As we have discovered modernly with the 2008 bank bailouts,¹²⁵ the subsequent "quantitative easing"¹²⁶ programs, and the 2020 global-lockdown helicopter money,¹²⁷ coordinated sovereign debasement is much harder to detect. It is certainly more difficult to defeat. Unfortunately for the medieval rulers, they could never quite figure this out. They were too busy fighting wars and trying to steal land from each other. This, of course, got in the way of them coordinating their monetary debasement. Be that as it may, by the mid-to-late 1800s these systems had generally stabilized. Gold and bi-metal (gold and silver) monetary systems all roughly adhered to an international gold standard¹²⁸ by the late-19th century. This loose monetary coalition continued in fits and starts right up to the Great Depression. By World War II, most of the Western economies abandoned gold as a primary medium of exchange in favor of paper notes.

CHAPTER EIGHT: FIAT MONEY

As mentioned above, the idea and preferential adoption of fiat[129] (paper) currencies in lieu of gold gathered steam around the time of the Great Depression. Despite claims to the contrary, there were many problems with the gold standard[130] era. Nations would often try to leverage advantage by manipulating gold markets. Countries like Weimar Germany had massive bank runs. So while the gold standard was a largely functional system for international trade, it was far from perfect. As I think Lyn Alden correctly identifies in *Broken Money*,[131] with the advent of modern technologies like the telegraph, the ability to transact without settlement bloomed. This also served to create an ever growing imbalance between gold reserves and gold claims. It also helped facilitate that market tinkering noted above. At any rate, in the post-World War Two era, the Bretton-Woods[132] agreement created a new gold "peg" scheme. Bretton-Woods launched a system where United States dollars were convertible to gold at a statutorily fixed rate. The United States dollar was chosen for this role for entirely practical reasons. Bretton-Woods was signed in the wake of World War Two.

At that time, the United States held in her vaults approximately 80% of the world's gold reserves. The United States was also the only industrial nation that retained any significant manufacturing capacity. The European nations and their factories were little more than rubble after the fighting in WWII. This was not an ideal arrangement for any of the nations of the world at the time. They all intuitively recognized the United States was not exactly the most stable country in the world. Keep in mind, the U.S. had just risen from a massive civil war only sixty-years prior. The U.S. banking system was world renowned for its banking scandals and financial crises.[133] It had also only just emerged from the chaos of the "wildcat banking" era by the time Bretton-Woods was signed. But the United States was the best the world had at the time. So, that was what they went with.

Of course, the United States' promise to the world that it would overcome its marginal track record in banking and finance was hardly a secure transaction. In fact, it did nothing to keep the United States from continuing on its well-trodden historical path. This was readily apparent to anyone bothering to look. The United States balance of accounts became increasingly lopsided through the 1950s. By the time the issue of excess American dollar float versus American gold reserves available for redemption boiled to crisis proportions in the late 1960s, the Bretton-Woods arrangement simply imploded.[134] It only took a scant twenty-five years for the United States to completely renege on promises made post-WWII. President "Tricky Dick"[135] Nixon closed the gold redemption window in 1971 by executive fiat.[136] The U.S. assured the world it was a temporary measure. They continued to promise a return to dollar convertibility right up to the moment they formally abandoned the Bretton-Woods agreement in 1976 with the Jamaica Accords.[137]

Since then, all major fiat currencies in circulation today "float" their respective values. This market for currency is based on a complex system of international exchange.[138] It must be said here that the United States enjoys a privileged position in this exchange hierarchy. The majority of global trade is still, to this day, settled and accounted[139] for in U.S. dollars. This is largely because of Bretton-Woods and the enormous amount of economic momentum embodied in the U.S. dollar. Put another way, it is not that the world necessarily wanted to continue accepting fiat U.S. dollars after the U.S. refused to honor their commitment to redeem for gold at a fixed-rate. It was much more a function of maintaining the status quo. By the time Nixon closed the gold window, an entire global financial system was built on the U.S. dollar. I would be remiss to not point out the widespread adoption of floating fiat currencies in the industrialized West has also coincided with a rather unprecedented peace—at least in the industrialized West. I say this is a coincidence, because many an economist will point to this and call it a "win" for fiat currency. The base reality is more appropriately attributable to the momentum mentioned above.

The threat of mutual assured destruction[140] and the United States' insistence—by force or threat of force—to maintain global adherence to the U.S. dollar's pre-eminent role in trade and settlement certainly helps. While many today attribute a gold like status to the so-called "petro dollar,"[141] this is an incorrect assessment of that relationship. The agreement with Saudi Arabia was reached after a good bit of saber rattling by the United States. After the OPEC oil embargo[142] of 1973 nearly crippled the U.S. economy, lawmakers began commissioning Congressional feasibility[143] studies on capturing oil fields as a military objective. Being the largest commodity producer in OPEC, the Saudis very much blinked first. The agreement involved a lot of political maneuvering on both sides, with echoes of the Israeli "Six Day War"[144] rattling through the respective halls of power in the U.S. and Saudi Arabia. It was more of a military aid package in response to commodity producing nations challenging U.S. hegemony than any attempt to peg the dollar to oil. It also marked the acceleration of the Memorandum of Understanding[145] (MoU) age for the military-industrial complex. Writ large, this all placed much woe, and warning, upon the militarily weak[146] commodity producing nations of the world. For them, it is widely understood, the U.S. is the one who knocks.[147]

Regardless, floating fiat currencies offer a number of advantages to the sovereign (government). Prime among them is the sovereign technically cannot go bankrupt anymore.[148] Trouble being, if the sovereign is irresponsible in the management of their internal finances, their ability to repay foreign debts plummets. This has occurred at various times and in various countries. The German Weimar Republic[149] is an infamous and extreme example of a country not technically going bankrupt, but being bankrupt all the same. For the last sixty-years or so, that has not been the case for the United States. The enormously privileged position of the U.S. dollar allows the United States to be incredibly irresponsible internally. All the while, foreigners keep taking her dollars anyway. This is how the United States has created an economy that derives 70% of its Gross Domestic Product (GDP) from consumer spending.[150] Put simply, the U.S. has been able to spend far more than it makes for a very long time. The

United States accomplishes this much the same way a trust-fund baby might finance a lavish lifestyle by running up the family credit card.

The broad cause and effect loop is the United States can acquire goods from abroad by paying in dollars that are continually being debased to deficit spend at home. In this manner, the United States has been able to offload and externalize inflationary forces while increasing the material comforts of the people within her borders. It is this unique ability to outsource inflation from monetary debasement that is also the primary creator and driver of what the late Dr. Graeber termed "Bullshit Jobs."[151] This misalignment of incentives has resulted in a system where the short-term gain of outsourcing most internal domestic productivity trumps long-term considerations of national security, stability, and domestic harmony. The looming trouble the United States faces now is that the world has gotten tired of this little hustle. While the U.S. blusters about, the rest of the world is apparently, and perhaps resolutely, moving towards a world that negates the U.S. dollar's privileged status.[152] To return to the trust-fund baby analogy, the family is about to cut-up the spoiled brat's credit card. They are just trying to figure out how to do it without ruining the family's finances.

Returning to the matter at hand, provided the sovereign is responsible in the management of their internal spending, fiat regimes provide a lot of flexibility to finance projects and unexpected expenses, like wars or natural disaster responses. If the sovereign needs a little more money than they have, they simply create a little more money to spend. But if the sovereign inflates their currency too much, so that the "value" of the currency drops too fast, they simply create less. The reduction in the issuance rate allows monetary liquidity[153] to "dry up." This is the essence of all the monetary policy talk around interest rates that dominates the financial media every quarter. Now, recall from earlier in the chapter, where it is posited that money:

1. Facilitates the exchange of value in low or zero-trust situations; and/or
2. Allows for the bulk extraction of fractional value from the personal or collective industry of sovereign subjects through taxation.

Then reflect upon the fact that the world of today is highly complex. Global communication is instantaneous. Nearly every point in the world is accessible by air within a day. Goods, materials, and partial-to-complete components are produced and shipped globally. We exchange value with strangers all the time, both foreign and domestic. Perhaps ironically, the glue that holds this amazing, complex infrastructure together is ultimately trust. In the modern economy, fiat currency, and more specifically, the U.S. dollar stands in for that trust. The trust component is that in any given exchange facilitated by U.S. dollars, the U.S. dollar stands-in for the promise to return like value. For instance, I do not know the person who runs the Starbucks on the corner and they do not know me. The U.S. sovereign-backed dollar provides the contract between us. If you give me a coffee and I give you U.S. dollars in exchange, the U.S. government sort of guarantees[154] you will receive like value later on by exchanging those dollars with someone else. In truth, the U.S. sovereign is running a scheme where they continually inflate the supply of money. The goal they aim for is one where you (everyone) loses 2% per year of purchasing power.[155] Leaving that aside for the moment, I have to pay my taxes in U.S. dollars. Through that mechanism, the sovereign extracts a fractional value of my industry. The sovereign in turn demands—by threat of force—that those dollars be accepted for all debts public and private.[156] This is a very convenient arrangement for the United States. As you can see, they have essentially baked the "coin clipping" style debasement from days of old right into the U.S. monetary cake. They also get to force everyone to trade in their fiat U.S. dollars. Just like the medieval kings, the U.S. sovereign can then use those dollars to buy all kinds of neat stuff.[157] The best part is, they do not even need an ounce of gold[158] to do it. King William of Orange[159] would have been proud.

Stability and Fractional Reserve Banking

In terms of economic stability, Switzerland[160] stands out as an exemplar. Those who follow the link may notice the United States is pretty far down the list at number seventeen. That places the U.S. between Belgium and Singapore. Keep in mind, of course, the total population and land mass of both is roughly equivalent to New York City. Despite the popular narrative that the United States is the "most powerful economy in the world," the truth is the United States is rather poor in terms of economic stability. In fact, some have even gone so far as to argue the U.S. has become a banana republic.¹⁶¹ That may be a bit of a stretch. Of course, it may not be as well.

Recall, the Bretton-Woods agreement lasted almost twenty-five years before it collapsed under the weight of deficit spending. American adventures in Korea and Vietnam (among other things) created deficits that would have bankrupted the United States if it honored the Bretton-Woods gold peg. The floating fiat system has now lasted for about fifty-five years. It is functionally imploding¹⁶² as we speak. Prior to the U.S. abandoning Bretton-Woods, gold was redeemable at a statutorily fixed rate of $35 per ounce. After the Nixon administration allowed the U.S. dollar to "float" on the open market, the result was a massive inflationary spike in the United States. Once the gold peg was broken, the world dumped their debased dollars *en masse*. By 1980, the same ounce of gold cost $800. Around the same time, then President James E. Carter Jr. made his (in)famous speech about a "crisis of confidence"¹⁶³ in America. For those living in Europe during the financial crisis of 2008, the speech has a lot of familiar themes. Key among them was the need for austerity measures¹⁶⁴ to "right the ship." In fairness to Mr. Carter, what he proposed was essentially correct. Unfortunately, the first big-R republican came on the scene in the form of Ronald Reagan. President Reagan's economic plan was to break out the national credit card via deficit spending. To understand this, you first have to understand the basics of central bank monetary policy. Without doing a complete deep dive, the simple version is the central bank wants to roughly balance the number of dollars going out vs. the number of dollars coming in with a catch: The number of dollars going out should be *slightly* higher than the dollars coming in.

The stated goal of the Federal Reserve is to keep this rate around 2% per year. The way they try to achieve this is by altering the interest rate they charge to banks either up or down. What Reagan did (and every administration since) was make the dollars going out far exceed the dollars coming in. The way they all achieve this is through monetary expansion,¹⁶⁵ i.e., "money printing." These images are from a paper¹⁶⁶ I published on student loan bankruptcies several years back. I use it here because it illustrates well what the result of that monetary debasement looks like:

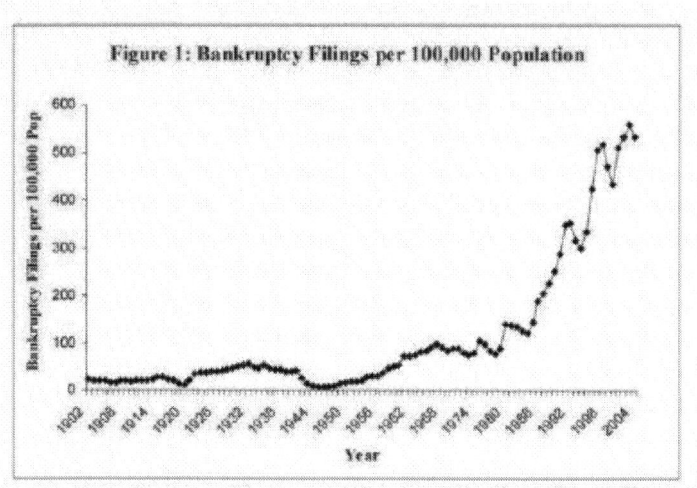

Figure 1: Bankruptcy Filings per 100,000 Population

Source: *Annual Report of the Attorney General of the United States (Through 1939) and Administrative Office of the United States Courts.*

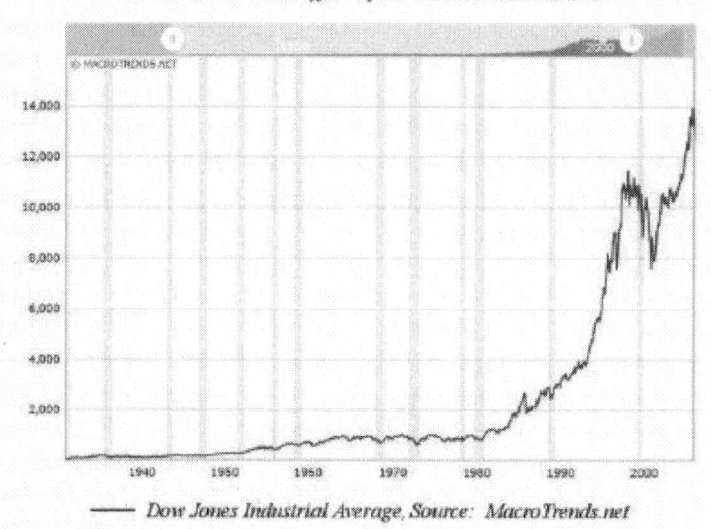

— Dow Jones Industrial Average, Source: *MacroTrends.net*

As you can see, starting right around 1980, there is a literal explosion of growth in both the Dow Jones Industrial Average and the rate of bankruptcies. If you look closer, you can see they almost track each other perfectly. What changed during this time? Massive deficit spending under the Reagan administration. Where Carter called for savings and austerity, Reagan threw a party and charged it to the national credit card. The result was a huge increase in credit issuance, a corresponding increase in debt defaults, and the on-going illusion of prosperity.[167]

To illustrate how this all works, lower interest rates roughly equate to lower borrowing costs. This, in turn, roughly translates into more lending. This is a key point, because modern money is created every time a bank lends money.[168] The way this happens is attributable to the idea of fractional reserve banking.[169] With fractional reserve banking, a bank only has to hold a certain percentage of the funds it loans out. As an example, imagine your father loaned you $100 at 5% interest. Then imagine you turned around and loaned that same $100 to your friend for 7% interest. Not a bad deal for everyone involved, assuming your friend pays you back. Let us change this hypothetical a little bit and add the "fractional reserve" part in. Same scenario, your father loans you $100 at 5% interest. Except this time you are allowed to loan that $100 to nine friends. But Father insists that you keep the original $100 in your pocket. The way you accomplish this is by giving each one of your nine borrower friends a card that has a $100 limit on it. They can then use that card to spend just like cash. The really neat thing is, they all have to pay you back $100 each at 7% interest. This is a pretty good deal, because if they all pay you back, you collect interest on loans totaling $900 *and* you will collect $900 over time. This is pretty sweet, considering you only have $100 in your pocket. In this "Father, Friends, and You" economy, you just created $900 of "new" money out of thin air.

This runs on the core idea that, as long as you lend money to people that pay you back, you will profit handsomely. Moreover, as the theory goes, more economic "energy" will be created. Of course, depending on the prudence of the lenders, that "energy" could be a wise investment. It also may very well be a boondoggle. With that in mind, if one of the borrowers defaults entirely and does not pay you back, you can still pay back the original debt to your father. In this simplified version, you have to keep a *fraction* (9/10) of the money you lend out in *reserve* in case a borrower defaults. For the banks, they have a lot of depositors that are (theoretically) earning interest. These are called demand deposits.[170] A commercial or retail bank must keep a certain percentage of those demand deposits available for withdrawal. The idea is, as long as everyone does not demand their money at once, the percentage held in reserve should cover any normal outflows. The rest can then be made available to lend out. Much like the nine-tenths reserve demanded by Father above, this was also roughly the reserve requirement the Federal Reserve had for banks in the

United States until 2020. Now banks in the U.S. have no reserve requirements[171] at all. Much like the Bank of England, the Federal Reserve relies on "prudence."[172]

If we expand this idea out further, what the Federal Reserve wants to do is ensure the "new money" being created through lending is roughly 2% more than the year before. This is where you here terms in the media about "target inflation," or inflation "running too high," or "running too low." Returning to the hypothetical "Father, Friends, and You" economy, the way Father would control how much you lend out, and to whom, is by raising or lowering the amount of interest he charges you in the first place. If he charges 20% interest, you have to charge your friends much more to borrow and still remain profitable. When it becomes more expensive to borrow, they will be less likely to want to do so. The same holds true in reverse. If Father only charges you 1% interest, then borrowing and lending become much less expensive. Thus, your friends will be more likely to borrow. Put more concretely, Father's base lending interest rate to you determines the cost of capital[173] to facilitate further lending to your friends. As touched on above, the ever present issue with this arrangement is the not-so-small problem of speculation. If it is very expensive to borrow money, your friends are much less likely to borrow to pay for very risky endeavors. If they want to continue to be able to borrow, they want to be sure whatever they are using the borrowed money to create will have a high likelihood of making more money than the loan costs.

But if it is very inexpensive to borrow, then it is much more favorable for them to take larger risks and seek larger rewards. This is a very simplified version of the "balance" a central bank is trying to achieve. They ideally want a rate that discourages naked speculation, but still encourages wise and thoughtful investment for growth. The trouble is, a handful of people in a room trying to decide what is, or is not a "good" investment and what rate to charge for that is simply a terrible idea.[174] To return to the "Father, Friends and You" economy, this logic presumes Father knows what is best for you and all your friends, without knowing who your friends are, what their skills and abilities may be, or what they are even planning to do with the money. This also leaves aside the niggling problem of you getting to "make" money from nothing save for a minor default risk. If we take these same problems and scale them up, they are magnified by the fact that

American banking has *never* been stable or reliable. That is roughly how we get to where we are today. Which then begs another question, "why is American banking so bad?"

American Banking

Those who watch mainstream news like Fox or CNN might lean into the idea that the divisions in the U.S. are getting worse each passing day. It makes sense why it seems this way. The deeper truth is many of these divisions have been in the U.S. since day one. To understand why, we must look briefly at what the U.S. really is, from a historical perspective. When the original thirteen states of the U.S. were established, they were all intended to be sovereign unto themselves. They were, in effect and structure, "new" nation states. The inhabitants of New York considered themselves to be citizens of the country of New York. Same was true for Virginians, Georgians, and all the rest—much like how we view the nations of Europe today. What drove the concept of the United States was a loose economic and political union of individual nations, similar to what the European Union is doing now. Some of those early "countries" in the U.S., especially in the slave-holding rural south, were particularly keen to maintain strong sovereign independence from the Union. When you hear arguments based on states rights,[175] this is where the idea comes from. The core of that idea envisions a relatively weak central (federal) authority and strong state authority. A poor, but roughly suitable comparison might be the United Kingdom in relation to the European Union. In essence, with "Brexit"[176] the U.K. did to the E.U. what the Confederate states tried to do to the U.S. during the Civil War.

On the other side is the concept of federal supremacy, which envisions a powerful central authority and limited state authority. While the E.U. is technically a confederation,[177] rather than a true federation[178] like the U.S., it shares many characteristics. Likewise, some of the same divisions and disagreements you see in the U.S. federal system are starting to crop up in E.U. confederate politics. Much as in the U.S., the political divisions in the E.U. tend to follow similar U.S. Liberal/Conservative ideologies. In turn, these ideologies are deeply rooted in historical concentrations[179] of economic activity: Urban-centered industrialism (liberals) and rural agrarianism (conservatives). For many reasons, and only generally speaking, urban industrialists tend to dominate the political economy[180] of a given nation. This is usually to the detriment of the rural agrarians. Not always true, but certainly more often than not. That calculus changed in the rural U.S. south in the early 18th century. Owing in large part to slave labor and the cash intensive crops of cotton and tobacco, the rural southern U.S. became an economic powerhouse.[181] To make a modern market analogy, the industrialized cities of the late 17th century were like General Electric, DuPont and IBM. In contrast, the rural agriculturists were like a huge swath of tech companies. Slaves, cotton, and tobacco were to the southern U.S. like what Microsoft, Apple and Amazon became in the tech economy. How does this all relate to banking? Strong central banks in the form of the First and Second Banks of the United States were tried, but ultimately failed. These failures were largely due to resistance and obstruction[182] by the agricultural south. A strong central bank grants an enormous amount of financial power to the actors that control the central bank. For example, one of the defining characteristics of the U.K. during its time in the E.U. was its refusal to adopt the Euro.[183] The United Kingdom kept the Pound Sterling. This was done to ensure the U.K. central bank retained its power to unilaterally set monetary policy internally.

Be that as it may, given the historical financial exploitation of rural agrarians, the U.S. south was highly suspicious of a strong federal bank. They wanted to retain control over internal monetary policy, much as the U.K. did prior to "Brexit." This resistance to central banking led to long periods of volatility and instability in the United States. The lack of a national central bank helped usher in the era of "wildcat banking."[184] With the wildcat banks came hundreds of currencies being used throughout the early United States. For a modern reference point, early U.S. banking looked very similar to cryptocurrency markets today. It was all over the place. You never knew if the "money" you held one day would be valuable or worthless the next. Likewise, similar to how one cryptocurrency can be used in a particular ecosystem but be useless in another, much of the currency that floated around during this time had the same problems. If you had money from a bank in Philadelphia, it would probably work in Pittsburgh. But it was much less likely to be accepted in New York. Go a little further out and it might be completely useless in Boston. For the most part, this was the steady state of American banking in the post-Civil War era, through the Industrial Revolution and into the early 1900s. The system was extraordinarily fragile. There were a number of recessions, depressions, panics and financial shocks[185] throughout this period. To be fair, free banking[186] regimes throughout the world were not universally as convoluted and fragile as what was found in the U.S. As author Lyn Alden correctly points out, beyond the confines of the United States, the track record on global free banking is "mixed."[187] The broad takeaway to this point should be that monetary systems and the issuance of currency inevitably create intractable problems. For these intractable problems few, if any, have ever discovered sound or reasonable solutions that can sustainably withstand the inevitable forces of corruption.

It must be said here that these problems trace back to the origins of money. Modern economists and modern economic theories condition their core reasoning around the idea that money is a natural evolution of barter and trade. As described above, however, the nasty reality is that money, in the form we use today, has always been a tool of extortion and theft. Money is not, and has never been, the result of a natural evolutionary and progressive process. It did not simply emerge from free willed exchange to the free willed use of currency units issued by a sovereign. Indeed, many of these intractable problems may well stem from the fact that it is likely impossible for any sovereign to achieve a "fair" system. This is especially true if these systems are, as this book suggests, rooted in inherently unfair practices. Even so, the U.S. compromise solution to the chaos of the U.S. free banking era was the U.S. Federal Reserve Bank ("the Fed"). While the Fed has generally (and poorly) served to stabilize U.S. banking, the current regime is also still constrained,[188] and often befuddled, by many of the same issues and divisions present since the founding of the nation. As noted above, these problems are deeply rooted in an agricultural/industrial divide that predates the United States itself. None of this, of course, alleviates any of the core problems of sovereign money. As currently implemented today, it is an inherently suspect medium for the free-willed transaction of goods and services throughout an economy. Yet it is an error to discount the vast history of experimentation and refinement embodied within the current system. It is dangerous and counter-productive to assume that another system can simply manifest in its place. Any such notion will unquestionably fail to retain the benefits reaped from money thus far without leaving considerable destruction in its wake.

Which is to say, even if such a change could be conscientiously implemented, there would still be no guarantee that the next system would be any better. Indeed there is a very high risk that the next system may prove to be substantially worse. Those unfortunate countries that experimented with Communism post-WWII learned this lesson all too well. The human toll[189] extracted in pursuit of that revolutionary and seismic monetary policy shift is shocking to reflect upon. The sheer savagery[190] and enormous loss of life that followed the violent implementation of Communism should serve as a stark warning. Anyone wishing to upend the system we rely upon now, imperfect and unfair as it may be, risks much for very uncertain outcomes.

CHAPTER NINE: OTHER MONEY

A question worth answering at this point is, "why inflate the money supply at 2% per year?" While I suggest it is simply "coin clipping" on steroids, others say it is just an arbitrary number.[191] To be fair, the central banks of the world have done a fair amount of research[192] to support the idea of a 2% inflation target. Moreover, famous economists, such as Milton Friedman, have also argued in favor[193] of a relatively low, but continuous inflation rate. From a historical perspective, the gold supply inflates between 2–5% per year[194] from mining and extraction. Given that gold has functioned as money for hundreds, if not thousands of years, it would not strain logic for the economically inclined scholars of the world to assume such inflation is beneficial. It must be said, however, that much of the research done by the central banks is premised upon economic theories[195] that say declining prices (deflation) are a negative outcome. As the old saying goes, reasonable minds may differ. Suffice it to say, very reasonable minds very much differ[196] on that premise.

While this may come as a surprise, ultimately the artificial target rate of inflation is immaterial. Whether the target rate is set to 1% or 20%, the effects are the same. The chosen rate only moderates the intensity of those effects. Modern economists, deeply schooled in the Keynesian tradition,[197] will certainly disagree. They will prattle on about their models, maintaining "real" balances, labor quantity, and so on. All the while, and as detailed above, they ignore the foundational errors in the core assumptions of their field. Or, as the late David Graeber derisively termed those errors, the "founding myths" of the field of Economics. As a practical matter, the effects of target inflation noted above are demonstrable. In fact, they are readily ascertained by even the most casual observer. We need only look at what is best described as "other" money to see them. Let us start by returning to the Economics 101 description of money: *Medium of exchange*, *unit of account*, and *store of value*. Medium of exchange is simple enough. I give you one dollar, you give me one dollar's worth of stuff. Same for unit of account. You owe me $20 dollars. We know this because we both agreed to it and wrote it down. But what of store of value?

If, as suggested by prevailing inflationary economic theories that inflation is a good thing, how can a dollar "store" value if it is being debased by 2% per year? The trouble here is two-fold, provided one accepts as reasonable the quantity theory of money.[198] Put simply, the quantity theory of money says, all else being equal, adding more money to the system dilutes the purchasing power of that money. This is much the same concept as adding soda water to a soft-drink. The more soda water you add, the less the beverage will resemble a soft-drink. The same is true for money and your purchasing power. The more currency units they add to the system, the fewer things you will be able to buy with those currency units. The two-fold problem being:

1. The purchasing power of the currency is continually being debased.
2. In response, everyone in the economy must raise their prices to compensate for that loss of purchasing power.

Put simply, and in line with the beneficial inflation theory, purchasing power must continually decline, while prices must continually rise. What should be patently obvious from this arrangement is that holding currency is a sucker's game. In turn, this renders useless the store of value function of currency. Or, to put it into the Economics 101 definitional framework, one-third of the economists' money definition suddenly becomes moot. That is, of course, provided one only views money through the lens of currency. It is here we can begin the discussion of "other" money.

What follows is well summarized by a recent quote. During a live interview, billionaire tech entrepreneur Michael Saylor noted there is a term for people that store wealth in currency. He said, "We call them poor."[199] The same "sucker's game" sentiment is echoed above for the reasons stated there. To better understand what is meant here, we can engage in a brief thought experiment. To that end, I pose this question to you, "If you inherited $5 million, what would you do with the money?" I have presented this question to hundreds of people. Everyone from students of mine, to co-workers, to family members. Nearly all of the respondees were some form of wage earner. By and large they do not operate businesses. They are not investors. Just run-of-the-mill middle-class worker bees. When asked what they would do with a $5 million windfall, almost universally, they say they would:

1. Pay off debt; and
2. Buy something.

The things they dream of buying run the gamut of tastes. A boat, designer clothes, a fancy house, a new car, entertainment systems, an exotic trip, etc. What I almost never hear is that they would buy an asset. In their mind, having a sufficient quantity of currency would enable them to acquire desirable things or experiences. A remote, and largely unrecognized second order effect for them is the relief from inflationary forces. They can all see and understand that the prices they pay for things are continually rising. In fact they often wax poetic about it. They will say things like, "I used to be able to buy a hamburger for $0.99, now they cost $5." Yet the forces that underlie those price increases are given scant consideration. Those price increases are simply accepted as a natural phenomenon. For them, increasing consumer prices are no different than the seasons or the tides.

This is one of the more insidious and pernicious effects of gradual monetary debasement. The purchasing power of their wages is continually being debased and vanishingly few[200] are able to increase their wages sufficiently to keep pace. This is scarcely noticed if the inflation rate is kept low. It becomes painfully obvious if the inflation rate suddenly accelerates. By contrast, however, if I pose the same, "What would you do with $5 million?" question to a wealthy person they respond very differently. Unlike their wage earning counterparts, they know that holding that cash is sub-optimal. They also know that spending it on consumables is a very poor use of that money. Instead, they almost universally choose assets. The pertinent definition[201] of an asset here is:

1. Something valuable belonging to a person or organization that can be used for the payment of debts.

In other words, money. Hence the chapter title, "Other Money." Based on my admittedly limited personal experience, most people do not consider non-currency assets to be money. For them money is only represented by cash or the number on their checking account balance. Yet, if questioned about the status of a figure like Elon Musk or Donald Trump, they will readily acknowledge those people are wealthy. What they apparently fail to realize is those men are, by and large, cash poor. They rarely carry or hold any significant sums of currency. The vast majority of their wealth is "contained" in assets. The wealthy understand that holding currency is a sucker's game. This is precisely why they do not do it. If you ask them, "how do you preserve purchasing power with currency?" Their very simple answer is always, "Buy assets."

The trouble with this reasoning is, the most commonly held assets do, in fact, depreciate or lose value. Take real estate, for example. Houses degrade with exposure to weather, normal use, and time. Equities (stocks) suffer from the same problem. If a company is not continually increasing revenue faster than monetary debasement, then that company is also going to depreciate. With that depreciation, their stock value will also decline. Art and collectibles are also subject to degradation over time. Even so-called "stable" assets like US Treasuries and bonds can rapidly depreciate[202] if monetary policy becomes too volatile.[203] The latter issue aside, the question we need to contemplate here is, "how does owning an asset overcome the store of value problem with currency?" The answer is embodied in a concept known as a "monetary premium."[204] As the name suggests, assets can and do accumulate a premium over, for lack of a better term, their "fair" value. Keeping in mind that all value is subjective, the idea here relies upon what might be best described as a rough, local consensus value. For example, a company's value is commonly estimated by its share price to earnings ratio[205] (P/E ratio). This simple calculation takes the amount of earnings in a given period, divides it by the number of equity shares outstanding and then divides that result by the market price per share. A monetary premium is realized when the nominal market price of the shares is higher than the earnings per share. For a frame of reference, the so-called "Magnificent Seven" stocks currently command an average P/E ratio[206] of

50. That ratio is double the average of the other 493 companies in the S&P500 index. This means that, on average, the dollar cost to buy a share in a "Magnificent Seven" company is 50 times higher than those companies' actual annual earnings divided per share.

What drives this monetary premium? Investors using stocks as "money." Or, more specifically, investors using stocks as a store of value that appreciates in nominal price roughly equal to or, ideally, superior to the rate of sovereign monetary debasement. Because there is an artificial scarcity to the issuance of equity shares, when more and more people use them as a store of value, the currency cost to acquire those shares goes up. Put simply, if more people want shares than are available, the price goes up. This results in a positive feedback loop for investors. They put money in now and sell for a higher price later. Easy peasy. Easy that is, provided the illusion of "forever growth" by the company is maintained. The principle way that illusion is sustained comes from inflationary monetary policy and the Cantillon Effect.[207] To boil it to pertinence, the Cantillon Effect says the closer one is to the source of monetary debasement, the more benefit they will accrue. This notion, again, relies upon acceptance of the quantity theory of money.[208] Provided you accept that premise, then the Cantillon Effect is readily explained. As new money is created, if you are the first to spend it, then the effects of monetary dilution are unevenly distributed and in your favor. At the moment of creation, the purchasing power has not yet been widely diluted. As that dilution works its way through the monetary system, the end recipients of the money are the only ones that feel the full effect of the debasement. Much like the soda water and soft-drink analogy above, if you pour soda water into the top of a large container of soft-drink and then immediately draw from the bottom, your beverage will still taste like a soft-drink. However, if you must wait to draw from the container, the more time the soda water has to reach solubility[209] with the soft-drink, the less the soft-drink will retain its flavor.

So it is with money and monetary debasement. The first in line to draw from the spigot is, of course, the sovereign. They can use those freshly debased dollars with the same purchasing power as the currency commands at the time of creation. The next tier-down are the banks and finance companies. Both of which acquire assets and lend with nearly the same purchasing power as the sovereign. Mega-corporations feed next. Then the large corporations and minor sovereigns, such as states, counties and municipalities get their turn. Below them lie the high wage earners. Lower still are the small businesses and government functionaries. Lastly are the growing ranks of the functionally and actually impoverished low- and non-wage earners. For the last fifty-to-sixty years, what the high wage earners and above have done is take their excess devaluing dollars and use them to purchase assets. Stocks and bonds have been the mainstay for nearly that entire period. The first 20-ish years also included using or starting small manufacturing, wholesale, retail or service businesses as assets. The decline of the desirability of those assets is directly attributable to the forces described above. As prices continually rise, while purchasing power continually declines, competitiveness in those commercial spaces becomes increasingly dependent on economies of scale.[210] Thus the relentless decline of the small business owner in favor of the mega-corporations like Wal-Mart and Amazon. In turn, this has led to the phenomena of the last twenty-to-thirty years, which has seen more and more dollars being devoted to real estate as an asset, especially residential real estate. In fact, this latter asset class has now become an institutional investment for everything from family offices,[211] to Real Estate Investment Trusts (REITs), to the asset management behemoths like BlackRock[212] and State Street. The most acute concentrations for the institutions are in multi-family and apartments, but there is an ever-expanding reach into single-family housing units as well.

Due to the relative, and oftentimes artificial[213] scarcity of these assets, as more and more investment dollars come in, the nominal price of these assets naturally rise. When you hear an investment advisor say things like, "The average return of the S&P500 is 10%," or "Real estate always goes up in value," the effect just described is where those ideas come from. What finance managers have historically tried to do is provide returns from those assets that exceed the rate of monetary debasement, i.e., inflation. The birth of the modern Exchange Traded Fund (ETF) came from research[214] that demonstrated funds managed by someone trying to beat the market, or indeed inflation, tend to underperform a broad index. In turn, if one bothers to look, it is patently obvious that the "returns" enjoyed by the S&P500 are highly correlated to the rate of monetary debasement.

THE GREAT REALIGNMENT

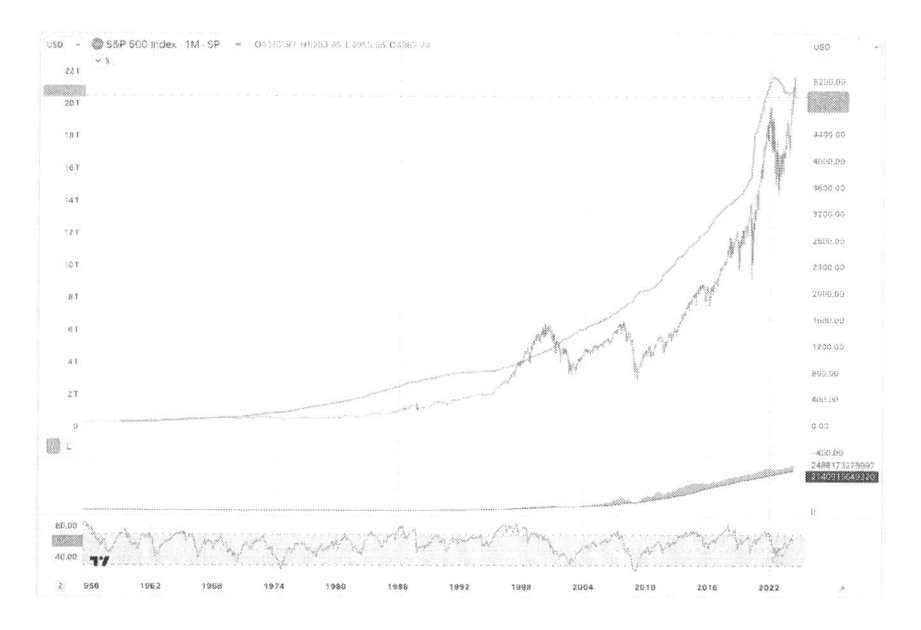

The chart above shows the M2 money supply[215] (also known as "broad" money, the smoother line) and the performance in dollar terms of the S&P500 index (the jagged line). This is very important to understand in the context of the current discussion. As you can see, if you had bought a single share of the S&P500 index in the mid-1990s, it would have cost you roughly $400. If you held that until today, you could sell that same share for nearly $5000. But if you had instead taken that $400 and put it under your mattress, that same $400 would only buy between $112 to $219[216] worth of "stuff" today.

In 1995, the relative **price** worth of **$400.00** from 2023 is:
$200.00 using the Consumer Price Index
$219.00 using the GDP deflator

In 1995, the relative **wage or income** worth of **$400.00** from 2023 is:
$171.00 using the unskilled wage
$187.00 using the Production Worker Compensation
$140.00 using the nominal GDP per capita

In 1995, the relative **output** worth of **$400.00** from 2023 is:
$112.00 using the relative share of GDP

Turn this the other way around, however, and you will see why the wealthy choose assets to preserve and increase their purchasing power over saving in cash. It will also very clearly demonstrate what a "monetary premium" looks like in practice. The $5000 price you could sell your "share" of the S&P500 today would have bought the equivalent of $1400 to $2740[217] worth of "stuff" in 1995. That is the power of the monetary premium, at least in the short-term. This is also the source of what is frequently cited as growing wealth inequality.[218] When Michael Saylor said, "We call them poor," this is why.

In 1995, the relative *price* worth of **$5,000.00** from 2023 is:
$2,500.00 using the Consumer Price Index
$2,740.00 using the GDP deflator
In 1995, the relative *wage or income* worth of **$5,000.00** from 2023 is:
$2,140.00 using the unskilled wage
$2,330.00 using the Production Worker Compensation
$1,760.00 using the nominal GDP per capita
In 1995, the relative *output* worth of **$5,000.00** from 2023 is:
$1,400.00 using the relative share of GDP

Let us pause for a moment here and consider a couple of important points. First, the returns on the S&P500 for the last decade are almost entirely driven by the so-called "Magnificent Seven."[219] Recall, the average P/E ratio of these companies is at 50. Put another way, the market value of these shares is 50 times the earnings per share outstanding. This means that the vast majority of the purchasing power "saved" by these assets is simply due to the monetary premium these assets currently enjoy. But they currently enjoy this massive monetary premium because of the Cantillon Effect described above. Nearly every entity from the high-wage earners up to the banks, finance companies, insurance companies, and pension funds are using stocks like this to preserve or increase their purchasing power. The only reason the wealthy do this is because the sovereign issued currency loses purchasing power every year due to monetary expansion. Thus, they are all "forced" to continually add to the monetary premium in assets like stocks. This buying pressure continually drives the nominal price of those stocks even higher. In turn, this causes things like the "Magnificent Seven" to rise to nominal share prices that far exceed any possibility of them being able to rationally account for them. Most investors are aware of this on some level. The way they traditionally protected themselves from this "irrational exuberance"[220] of markets was through a mechanism called hedging.[221]

Plainly speaking, a financial hedge is like simultaneously placing a big bet on Red, a big bet on Black, and a tiny bet on Green on a roulette wheel. If done well, as the idea goes, you will make money no matter the outcome. Obviously if such a strategy were sound, roulette wheels would have gone out of business long ago. The reality in financial circles is they try to use probabilities[222] to better align their hedges, with the *hope* of protecting their upside gains from any downside losses. If you have ever heard someone refer to a "60/40 portfolio,"[223] that is a rudimentary hedge allocation based on a 60% weighting in stocks and a 40% weighting in bonds. The idea being, stock performance is (probabilistically) inversely related to bond performance. Put simply, and as the theory goes, if stocks go down, bonds go up and vice-versa. It is a 60/40 split because over the last fifty or sixty years, stocks have generally outperformed bonds.

Given the monetary conditions in place since Nixon rug-pulled[224] the world in 1970, this was a fairly sound strategy. The Fed would juice the markets a little, stocks would go up, you make money. Once the Fed went a little too far, they would pull the punch bowl, stocks would go down, but your bonds would rally. It remained a sound strategy right up until the mid-to-late 1990s. With the increasing availability of personal and enterprise computers came a new-fangled way to hedge: Complex derivatives.[225] Derivative contracts[226] have been around for ages in agricultural commodities markets. In comparison to some of the more exotic derivatives employed by the finance sector, commodity derivatives are downright boring. The idea is simple: If you are a company that makes products from corn, (for instance) you obviously need to buy corn every year to keep making products. The trouble is, someone has to grow that corn. To make matters worse, growing things is not a guaranteed prospect. Nature is a fickle mistress. Nature will often come along and screw up things like growing and harvesting corn. These uncertainties make it very hard to plan for the future. On the flip side, the grower also has problems too. If all the growers have a bumper crop, then the price they can charge for that year might go way down. Likewise, the next year might see a drought for half the farmers. Lucky for the ones that got rain in that case. They get to charge a lot more that year. The problem on the grower's side is much like the one for the product manufacturer. How can they reliably plan for the future? Should they buy more seed or less? How many people should they hire? Is it time for a new tractor, or should they wait?

Derivative contracts are a way to smooth out those ups and downs. They are used to hedge risks against weather, plant diseases, poor yields, fluctuating prices, etc. Both sides of the trade can take out derivatives contracts against future risks. The contract issuer, in turn, can earn money for taking on the risk. The end result is a much smoother and more predictable trading experience for the producers and purchasers of commodities. Prior to the widespread use of computers, trying to use derivatives in the financial markets was nearly impossible. Unlike the much more limited range of problems that can occur with commodity producers and buyers, the range of problems that financial products can encounter are virtually unlimited. The advent of increasingly more powerful and affordable computers, coupled with advances in complex mathematical modeling from fields as diverse as quantum mechanics, to astrophysics, and biology gave rise to the field of quantitative finance.[227] From deep within that realm sprang forth the creators of, and the markets for financial derivatives[228] in the 1990s. Once created, the derivatives market quite literally exploded. These financial derivatives were allowed to trade "over the counter" (OTC), meaning without oversight. By 2002, as the market for derivatives ballooned, Warren Buffet famously said these instruments are, "Financial weapons of mass destruction." Meanwhile, the Fed chairman at the time, Alan Greenspan,[229] acknowledged that there were *some* risks with derivatives. But he went on to say that he was "[Q]uite confident that market participants will continue to increase their reliance on derivatives to unbundle risks and thereby enhance the process of wealth creation."

As everyone knows now, Mr. Buffet was right and Mr. Greenspan was terribly, terribly wrong. When the derivatives markets began to unwind in 2008, it was indeed financial mass destruction.[230] This brings us back around to the beginning of this discussion of "other" money. As we have demonstrated, assets like stocks, bonds, and real estate are used as money. They are used specifically as a store of value against monetary debasement. The inherent problem with this scheme are the associated and unmitigated risks. Holding depreciating assets with the sole expectation of an increasing monetary premium cannot go on forever. Once the monetary premium far exceeds any income value derived from the underlying enterprise, the monetary premium *will* come down. Whether the income comes from rental fees for houses, or earnings from a business is irrelevant. At some point, it becomes obvious[231] the monetary premium is not supported by the productive reality of the underlying instrument. At its core, the concept of a monetary premium relies upon the greater fool theory[232] to survive. In its simplest form, the greater fool theory says the price will always go up so long as someone else is willing to pay it. The trouble with this idea is, the higher the premium goes, the fools become fewer and farther between. At a certain threshold, the fools simply stop buying at all. In laymen's terms, this is known as a "bubble."[233] In years past, when a bubble like that popped, only the last fools were left holding the bag. In market terms, they would call this a correction. The underlying logic being, the market is correcting the nominal price of assets down to levels that can be reasonably justified based on earnings or income.

The trouble we have now was revealed by the 2008 Financial Crisis. No longer are the risks simply being held by the last fool. With financial derivatives, everyone in the world has been turned into the last fool. The expression "too big to fail" reflects this. Financial derivatives are now so deeply interwoven into every single risk asset outlined above, that the widespread decline of one asset, such as stocks or real estate plummeting in value, will lead to a cascade failure[234] of the rest. That is why the global central banks keep resorting to such extraordinary levels[235] of monetary debasement to keep the scheme alive. To make matters worse, right in the middle of all their financial juggling, they collectively ran into the 2020 global shutdown. For perspective, during the two-years of that debacle, the United States quadrupled[236] the M1 money supply. This is the source of the price inflation that so many are being trampled under currently. More troublingly, that quadrupling of the money supply is straining the resources the central banks have far beyond anything they have previously conceived. Doom and gloom aside, what I aim to demonstrate here is that "other" money is what has kept the global financial system afloat since the United States abandoned Bretton-Woods. Unfortunately today, "other" money has now become the greatest source of risk to the global financial system. If you trace the complex web of derivatives to their juicy center, what you will find is quite literally nothing. There is no solvent counter-party to the risk²³⁷ those derivatives are meant to hedge. The systemically important²³⁸ bank, Credit Suisse, discovered this the hard way in 2023. The details are ugly enough that they have decided to keep the reasons a secret for 50 years.²³⁹ The constant fear among the central bankers now is what they should do if the forces that took down Credit Suisse (and others)²⁴⁰ spiral out of control. Which means, the question we should all be asking ourselves right now is, "how do we fix it?" As I hope I have demonstrated to this point, the answer to that question is far from simple. That aside, with all of this fresh in mind, let us take a moment here and add in the "other" to our money definition from earlier. To answer, "what is money," we can now say money is something that:

1. Facilitates the exchange of value in low, or zero-trust situations; and/or
2. Allows for the bulk extraction of fractional value from the personal or collective industry of sovereign subjects through taxation; and/or
3. Is an asset that appreciates in nominal price at, or near, the same rate as the sovereign increases the quantity of the first two.

To put this in perspective, common folk like you and me primarily use number one. Sovereigns primarily use number two. Wealthy people primarily use number three. Unfortunately, as we have been discussing, all three are facing serious systemic issues. Those issues have few easy solutions.

Modern Money

U.S. banking and finance today are governed by a mix of state and federal authorities with widely varying degrees of regulatory power. Add to that significant changes²⁴¹ in the way banks and finance companies were allowed to structure themselves in the 1970s through the 1990s, and you get modern U.S. banking and finance. In terms of relative stability, you might recall from earlier, the U.S. financial system is anything but stable. A suitable and terribly dated analogy would be the experience of using Microsoft Windows 95. Once you think you have it figured out and working, it just crashes anyway. The well-trodden concept of a "boom-bust cycle" being a natural and unavoidable phenomena is rather attributable to central banking and the monetary policies²⁴² they create. These institutions and policies are invariably and demonstrably[243] the cause of the so-called "boom-bust" cycle.

As stated at the outset, discussions around money are opaque, difficult to unpack, and have widely divergent points of view. There are endless sources for reasoning, and endless sources of presumed authority. The point of this exercise is not to definitively state that what is presented here should be accepted as incontrovertible fact. It is rather to illustrate that commonly accepted wisdom and folklore about money is often deeply flawed. Put differently, some of what is presented here is most certainly iconoclastic. It is also inevitably at stark odds with accepted monetary orthodoxy. What is unmistakable are the natural and predictable consequences of allowing academics who have demonstrably poor starting assumptions to dominate discussions around money. From the days of medieval kings, sovereigns have sought to capture and plunder territories and resources. They also desperately strive to retain power once seized. Each of these endeavors requires significant amounts of money to accomplish. In pursuit of these goals, sovereigns have in the past, and continue to this day, to manipulate money to achieve those goals. What is required in all cases is inevitably more money than the sovereign treasury holds. Their very simple and insidious method of circumventing that limitation has always been to debase the money they force their population to trade with and pay taxes in.

The uncritical acceptance of common understandings around modern money and monetary systems is much akin to uncritical acceptance of the ancient understanding that the Sun revolves around the Earth. The fact that such information is dogmatically and thoughtlessly reinforced ought to be evidence enough to warrant caution and curiosity. This is especially true if one does not agree with the sovereign "right" to take productivity from their population beyond what is required for the protection of the people that reside within the sovereign's sphere of influence. Yet this is the visible result today. Those results have been catastrophic for all that run afoul of the powerful sovereign nations that wish to impose their will upon others. This chapter outlines the rough method by which those mechanisms are employed. This brief historical journey is not comprehensive. It is rather to illustrate the complexity and opacity of modern monetary policy. It is also to highlight the fact that the academics that lend support and comfort to those policies do so for a reason. Neither is because the inherent functions of money require complexity to operate. Rather, and in both cases, the complexity is a means to obscure the true purpose of the functions of money and monetary policy as employed and taught. Such obfuscation is very beneficial to those who wield the levers of power in finance, banking and international and domestic relations. In turn, those in power elevate and propagate the academic information that best helps entrench their position.

Yet simply gaining a deeper understanding and insight into this opaque field is far from adequate. It must be said with no uncertainty that much good has also come from this devilish arrangement. While it may be easier and more satisfying to bemoan the declining state of the world, the reality is quite different. While it may not be readily apparent, the quality of life and standards of living have risen significantly[244] across the globe since the advent of the modern monetary system. Such improvements are not uniform by any means. But, as a general rule, more people around the world are better off today than they were 50 years ago. The trouble is, these improvements have come at a cost. Much of that growth and innovation was essentially bought and paid for with debt. The natural and predictable consequence of this mass debt experiment, unconstrained by anything but imagination and the ability to maintain the illusion of solvency, has been two-fold:

1. A substantial increase in overall standards of living globally; and
2. An increase in wealth inequality,[245] that is slowly and regressively reducing standards of living globally.

Put simply, and to repurpose an old and discredited idea from the oil industry, we may well have reached "Peak fiat."[246] This is the unfortunate result of relying on debt and monetary expansion to achieve the aforementioned improvements in global standards of living. Improvement through debt and debasement can only result in the manifestation of the second problem of grotesque wealth inequality and subsequent decline. The larger and much more dangerously looming issue is when the debt from monetary expansion stops working.[247] It is much like if you ran up all your credit cards to improve your quality of life in the short-term. Sooner or later, that credit line will run out. Once it does, the quality of your life will decline rapidly thereafter. On a global scale, this declining process is blunted and masked by the sheer size of the global economy. But underneath, the same forces are at play globally as they are at the individual level. Sooner or later those forces will inexorably result in the same outcome globally—bankruptcy. Yet, much like the overextended credit card holder, the solution to the burgeoning global monetary debt problem requires much vigilance, conscientious planning, careful savings, and a quick return to sustainable spending practices. Such a path also relies heavily on luck.

For if you, as an overextended debtor, come upon catastrophe, it is all too easy to return to the credit card to borrow your way through the problem. The same is true at the sovereign level, except the sovereign has far more influences pushing towards the easy way out (monetary expansion), with few, if any, demanding the opposite. Strict austerity measures and carefully managed debt wind-down planning are toxic to the political class. Thus, they are, for all intents and purposes, frozen in amber.[248] Moreover, modern "sovereignty" is no longer just a king and his court. Today, the sovereign is a distributed system comprised of thousands of actors. Unlike an individual debtor, there is no one person responsible for the success or failure of the system. As such, factions develop, with each pulling in different directions. When monetary conditions grow tighter and more stark, the factions are not incentivized to work together. In fact, they are rather incentivized to do the complete opposite. Their survival depends upon their ability to grab as much as possible before the whole thing implodes. This would be much like ten debtors all sharing the same credit card, with all beholden to none but themselves. With no one to blame and no one to lead, the result is predictable with near certainty. In fact, it has been the same result[249] in every system that has expanded their money beyond their means to redeem the debt incurred. From individuals to the nations of the world, time and time again, for thousands of years, the end is always the same: Catastrophic collapse.

While this may seem dire and rightfully so, such an outcome is not pre-ordained. Such outcomes are a natural and predictable consequence of centrally controlled money, regardless of form. Thousands of years[250] of human history have clearly demonstrated this is an unpreventable outcome from the use of centrally controlled money. Whether we call it greed, hubris, arrogance, or a road to hell paved with good intentions, it is, in all cases, the central control of money that is ultimately driving these failure outcomes. But we must also acknowledge that money in and of itself is not "evil." And just as much as the current monetary system has wrought much evil on the world, it has also brought much benefit. The trick that no one in history has ever managed to figure out is how to avoid the former without sacrificing the latter. The modern system came close in some regards. By other measures, it has been absolutely abhorrent. But the abhorrence unleashed thus far may very well pale in comparison to the precipice financial derivatives have brought us to. Now that we are standing at the edge of this precipice, the burning question is, "Must we go over the edge?" I say confidently and resolutely that the answer to that question is, "No, we do not."

PART THREE: GREED

CHAPTER TEN

Unlike power and money, greed is not an abstract concept. It is not a human designed and created system. Lao-Tzu,[251] the 6th Century B.C. Chinese philosopher said, "There is no greater calamity than greed. There is no greater disaster than not knowing contentment. There is no greater fault than avarice." Greed finds itself counted among the Seven Deadly Sins,[252] coming in at number two after pride. Some American Native tribes considered greed a form of psychosis[253] and were (and are) quick to associate that psychosis with the European-Christians. The modern English definition[254] of greed says, "A selfish and excessive desire for more of something (such as money) than is needed."

Indeed, one would be forgiven if they assumed greed is universally reviled and condemned. Yet the Objectivist-Libertarian philosopher Ayn Rand may not have agreed. In her 1966 treatise, *Capitalism: The Unknown Ideal*,[255] Rand says, "Since time immemorial and pre-industrial, 'greed' has been the accusation hurled at the rich by the concrete-bound illiterates who were unable to conceive of the source of wealth or of the motivation of those who produce it." In this, it appears Ms. Rand instead argues that the accusation of "greed" is an ignorant misunderstanding of the motivations and intentions of the productive class. While separated by decades, Kate Luckie, a medicine woman of the Wintu Nation, prophesied the following to Cora DuBois in *Wintu Ethnography*:[256]

When the Indians all die, then God will let the water come down from the north. Everyone will drown. That is because the white people never cared for the land or deer or bear.

When we Indians kill meat, we eat it all up. When we dig roots, we make little holes. When we build houses, we make little holes. When we burn grass for grasshoppers, we don't ruin things. We shake down acorns and pine nuts. We don't chop down the trees. We only use dead wood.

But the white people plow up the ground, pull up the trees, kill everything. The tree says, "Don't. I am sore. Don't hurt me." But they chop it down and cut it up. The spirit of the land hates them. They blast out trees and stir it up to its depths. They saw up the trees. That hurts them.

The Indians never hurt anything, but white people destroy all. They blast rocks and scatter them on the ground. The rock says, "Don't. You are hurting me." But the white people pay no attention.

When the Indians use rocks, they take little round ones for their cooking. The white people dig deep long tunnels. They make roads. They dig as much as they wish. They don't care how much the ground cries out. How can the spirit of the Earth like the white man?

That is why God will upset the world—because it is sore all over. Everywhere the white man has touched it, it is sore. It looks sick. So it gets even by killing him when he blasts.

But eventually the water will come.

Taken in the context of Ayn Rand's approximation of greed, it appears the American Natives were not ignorantly railing against productivity while clamoring for their share. It would rather seem they railed against callous waste and wasteful practices. Which all leaves us in a bit of a conundrum when it comes to the notion of greed. Is greed immoral? Is it just misunderstood? Is it rational self-interest? Or is greed truly a form of psychosis? When we invoke the idea of greed, we collectively do ourselves a disservice. To borrow from one of investor Charlie Munger's famous quotes,[257] what we really ought to be focusing on are the incentives. Greed is a problematic concept for two big reasons:

1. Greed is a highly imprecise metric that relies upon an even more imprecise moral standard to evaluate; and
2. The very concept of greed violates Hanlon's Razor.[258]

For those unfamiliar (or who have forgotten from earlier chapters), Hanlon's Razor says, "Never attribute to malice that which can be adequately explained by incompetence or stupidity." And if we were to try to reconcile the statements in the previous chapter by Ayn Rand and Kate Luckie, I think Hanlon's Razor is best situated to do so. Ayn Rand is correct—malicious greed is not necessarily the driving force behind someone like Elon Musk, or indeed, the great oil and rail barons of bygone days. Likewise, Kate Luckie is correct—waste and destruction unquestioningly feature prominently in competitive resource capture economies. And, unquestioningly, Elon Musk and the great oil and rail barons of bygone days have created unfathomable waste and destruction in pursuit of technological advancement.

So, if this does not speak to malice, then where does the incompetence and stupidity of Hanlon's Razor lie? The answer is found within the structure of the game itself and the rules that naturally follow. To this point in the book, I have done my best to refrain from commenting too extensively on the current political economy we collectively find ourselves operating in. This is a deliberate choice, as I very much want to "prime the pump" so to speak. As I have mentioned more than a few times, the historical record of the last two parts is necessarily condensed to what might be loosely regarded as "first principles." That term is bandied about rather freely these days. I find that those who most reference first principles seldomly invoke them correctly. I only invoke them hesitantly here.

In the context of this book in particular, greed takes its place among the human systems of power and money for a specific reason: The evaluation of incentives. More specifically, we explore the idea of greed to further and better understand how the in-built incentive structures of power and money can, and often do, lead to negative outcomes, like resource hoarding and waste. But it is also to unpack and explore how the incentive structures can, and often do, lead to highly positive outcomes, like human flourishing and technological advancement. Moreover, the discussion around greed will also (hopefully) lead nicely into the discussion about Bitcoin and how this all fits together.

Incentives

The Charlie Munger quote alluded to above says,[259] "Never, ever, think about something else when you should be thinking about the power of incentives." I think Mr. Munger was correct. As noted in the Introduction, this book is intended to challenge common assumptions about the modern political economy and how it relates to the invention and adoption of Bitcoin. The first two parts of the book have hopefully provided a richer context and understanding of the evolution of money and power than is customary. If you have made it this far, then perhaps some of it has resonated. Now that we are embarking on the discussion around greed, it seems appropriate to bring the conversation more concretely into the particulars of the modern world. In pursuit of this, let us turn for a moment to the dictionary definition[260] of greed, which says, "A selfish and excessive desire for more of something (such as money) than is needed." Seems reasonable enough, does it not?

Yet, without poking too far, we run into a few problems. For instance, what does it mean to be selfish? Turning to our handy dictionary,[261] it says in relevant part, "Concerned excessively or exclusively with oneself: seeking or concentrating on one's own advantage, pleasure, or well-being without regard for others." Okay, fair enough. But what of "excessive?" Again we turn to the fine folks at Merriam-Webster who say excessive is[262] that which is, "Exceeding what is usual, proper, necessary, or normal." If only this were sufficient, we could press on. But, alas! It is not. Our last niggling little definitional problem with greed is what do we mean when we say something is "needed?" With one last turn to the dictionary, we find that need is[263] when, among other things, there is, "A lack of something requisite, desirable, useful" and "A physiological or psychological requirement for the well-being of an organism." So, when we put this all together then, being greedy means:

1. Seeking or taking advantage for oneself without regard for others

2. By taking more than is usual or "normal"
3. Beyond that which is desirable, useful and/or is physically or psychologically required for well-being.

Easy enough. And, with that in mind, I pose to you the question: Is Elon Musk greedy? Does Elon Musk seek or take advantage for himself without regard for others? It depends on your perspective, right? Likewise, does he take more than is usual or normal? Leaving aside the problem of what "normal" taking means, Elon Musk is currently worth over $200B. Since that is a big number, here is a fun visual. With credit to Humphrey Yang on YouTube,[264] this is what Elon Musk's net-worth looks like if a grain of rice was equivalent to $100,000:

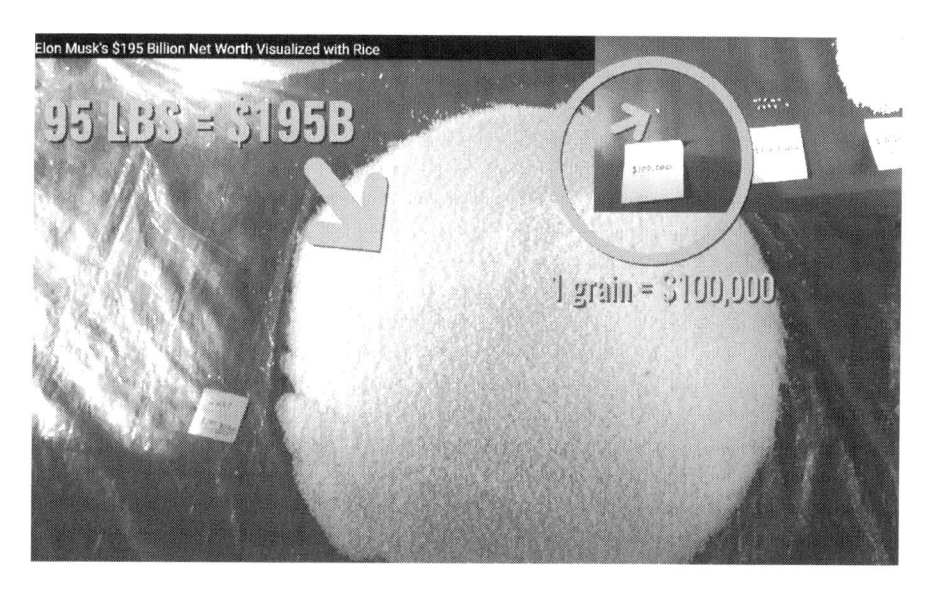

I am not sure what that looks like from your perspective, but it certainly appears to me that Mr. Musk has "more than is usual or normal," does he not? Take it a step further and we must ask, is this "desirable, useful, or required for Musk's psychological or physiological well-being?" Maybe.

There are plausible arguments to be made in any number of directions. Depending on who you ask and what their particular political, philosophical and ideological leanings may be, given what we know so far, Elon Musk may, or may not be, "greedy." Right? So, while we are on the topic, let us unpack some of Elon Musk's motivations for acquiring so much wealth. In a recent court battle[265] over a proposed compensation package for Mr. Musk's service at Tesla, the court noted:

> *Musk is motivated by ambitious goals, the loftiest of which is to save humanity. Musk fears that artificial intelligence could either reduce humanity to "the equivalent of a house cat" or wipe out the human race entirely.*
>
> *Musk views space colonization as a means to save humanity from this existential threat. Musk seeks to make life "multiplanetary" by colonizing Mars. Reasonable minds can debate the virtues and consequences of longtermist beliefs like those held by Musk, but they are not on trial.*
>
> ***What is relevant here is that Musk genuinely holds those beliefs.***
>
> *Colonizing Mars is an expensive endeavor.* ***Musk believes he has a moral obligation to direct his wealth toward that goal***, *and Musk views his compensation from Tesla as a means of bankrolling that mission.*

The **Bolded** text is my addition to the judge's writing. Keep in mind, this is not dicta.[266] This is in the published trial record, which was written by a very capable and competent legal fact finder.[267] Meaning, as a purely legal matter, it is a fact that Elon Musk genuinely believes and thinks it is a moral duty to save humanity from an existential threat by colonizing Mars. To achieve this goal, Musk is apparently doing everything he can to acquire massive amounts of wealth to fund this endeavor. Moreover, in a citation to the quoted text above, the court also said:

> *Musk does not dally in the conventional amenities of ordinary billionaires. For example, he owns only one home. [Citation omitted] ("I tried to put it on Airbnb, but they banned Airbnb in Hillsborough. They're so uptight.").*

The point being, someone of good conscience and moral character could very easily look at the enormous wealth and power that Elon Musk has acquired and call it greedy. They would not be wrong. Likewise, another person could look at him and say the complete opposite. They would not be wrong either. This is especially true if they might happen to share Mr. Musk's rather dim prognostication on the likelihood of humanity's continued existence after artificial intelligence relegates us to the status of a house cat. Worse yet, a third person could come along and assume Mr. Musk is completely full of beans and argue Musk knows that artificial intelligence would never do such a terrible thing. Based on that evaluation of Musk, they too could conclude Musk is either greedy or not, just depending on how they subjectively view any number of other metrics they might hold dear.

Since Elon Musk is the only person on Earth that can reliably tell us what his motivations are, we can only guess. That said, we could also return to Charlie Munger's view of the world and look at Musk's incentives and gauge everything he has done through that lens. For example, there are credible arguments[268] that say electric cars and their batteries are actually more harmful to the environment and humanity writ large than cars burning fossil fuels. In fact, there are credible arguments made that say the entire "Green Revolution"[269] is completely unsustainable. If taken as true, those arguments say that, in fact, the Green Revolution is a rapidly emerging environmental catastrophe, completely contrary to the stated goals. Elon Musk is a highly sophisticated, well-educated person. Is it possible he is just not aware of these problems? I think that is very unlikely. That said, is it possible that Musk is aware climate change concerns command significant attention, government grants and assistance, and is likely a strong source of capital? I think that is very likely.

So, let us put that all together as a little thought experiment. To start, assume you are a person that knows humanity is at risk and the only way to save humanity is by colonizing Mars. You know that is really expensive to do. Knowing both those things, would you create a company that you know is harmful to the environment, does nothing to mitigate climate change, but you think is very likely to receive massive investments and make enough money to fund that dream? And, if so, what is your true incentive? To save an environment that is probably doomed anyway? Or is your incentive to save humanity? Put another way, is it unethical to raise money for something false in pursuit of something noble? What if there was no other feasible way to fund the noble purpose? Does that make the falsehood okay? Regardless of your opinion, this still leaves out another problem: What if your noble purpose is batshit crazy?

With all this in mind, is it also possible that a particularly nasty and self-serving person could lie about both helping the environment AND saving humanity in order to create a company that could potentially make them the richest person in the world? Would it not be possible that someone could lie about two noble purposes in order to convince people they are ethical and noble, when in reality all they care about is acquiring wealth and tricking people into giving it to them? What would Charlie Munger say the incentive was? Lust for power? And, if so, how would he know if this person is legitimate, a con-man, a lunatic, or a psychopath? It gets tricky, does it not?

This highlights a very important point in this book. One of the major issues we come across in a competitive, resource-capture based economy is the problem of hidden information. In a competitive environment, regardless of whether it is football, poker, chess, or indeed, business and politics, it is always advantageous to hide information from your opponent, especially about your motivations. This is a regular feature of advertising and marketing, sales, mergers, acquisitions, electioneering, dating—the list is virtually endless. There is a popular quote, which is incorrectly attributed[270] to Sinclair Lewis, that says, "When fascism comes to America, it will be wrapped in the Flag and carrying a cross." The authorship is irrelevant. The underlying meaning is what is pertinent here. What the quote is really saying is, fascism will pretend to be pro-America and pro-Christian in order to trick American people into accepting it.

In other words, because fascists know they will be resisted, they must resort to trickery and deception to seize power. With that in mind, we could also, potentially, say the same for Elon Musk. If Elon Musk wants to become the richest and most powerful man in the world, simply telling everyone that is what he wants and demanding that it happen would not work. Let us pretend for a moment that, in the dark recesses of Elon Musk's mind, he is actually a Bond Villain.[271] He desires nothing more than to wield absolute control and authority over every human being on planet earth. In pursuit of this goal, he decides the way to achieve global domination is by launching a constellation of satellites that completely encircle the planet. After he has his "Iron Dome" of satellites deployed, he plans to use advanced AI and powerful lasers to target all critical infrastructure on earth. Then, he makes his demand:

You must pay me $1 million or the world will end in a fiery hell!

— **Evil Doctor Elon Musk**

But before he can pull this off, the evil Dr. Musk has a few problems. First, that is going to be expensive. Those fricking satellites with laser beams attached to them do not come cheap. Second, people might get fussy about satellites with laser beams attached to them pointed at their heads. Third, maybe the good people of the world do not want a despotic, tech-obsessed leader holding them hostage for $1 million. What is an evil genius to do? To get his plan for world domination on track, evil Dr. Musk must first raise some money. Since he is an evil super-genius, he decides on a Master Plan. The Master Plan will ambitiously leverage "green" money to bankroll a car company, simultaneously invest in an AI startup, and launch (no pun intended) a space exploration company.

His evil genius sleight-of-hand will be to parade himself as a champion of the environment, a champion of open access AI, and a champion of a free and open internet. But, while doing this, he will simultaneously parade himself as a savior of humanity. "We must go to Mars!" will be his rallying cry. Instead of buying expensive homes and luxury cars, he will instead live in a pre-fab tiny home.[272] His evil, super-genius mind knows the midwit earthlings he despises, drunk on decades of science fiction stories and lives of leisure, will eat this stuff up. If he sells it right and makes it all just bold enough to be possible, then while no one is paying attention, he can bring his evil plan to fruition. The best part? These moronic earthlings will actually *pay him* to enslave themselves! Brilliant! (Insert evil laugh).

If this were all true, then it would be very easy to characterize Elon Musk as evil, greedy, self-serving and maniacal. The problem is, the only one that can truly know if it is true or not is Elon Musk. The point is, if he is an evil super-genius, then chances are good he will lie to hide those evil intentions. Or we can just take him at his word, believe he is good, that his intentions are good, and he really is just a boy with a dream of saving humanity. Moreover, there still remains the distinct possibility Musk is a moron. The likelihood that human beings can become "multiplanetary" during Musk's lifetime are functionally close to zero.[273] Meaning, just because Elon Musk is a great salesman and engineer, it does not necessarily follow that he can overcome the major obstacles to living on Mars. Free market champions may blather about some invisible hand[274] and the wisdom of markets, but free markets have never had a filter for bullshit.[275]

The point of all this being, it does not matter if he is being greedy or not. No matter how you shake it all out, Mr. Musk is rapidly accumulating enormous wealth and power. He already has a constellation of satellites[276] surrounding the earth. Partly because of that and partly because he has a fleet of escape velocity transport vehicles, he is also becoming deeply embedded with the U.S. military-industrial complex.[277] Likewise, this satellite array also deeply embeds him in global telecommunications. This is all true without even mentioning that, with his acquisition of Twitter,[278] he is now deeply embedded in global social media as well. Let us also not forget, Mr. Musk has his toes dipped in public infrastructure[279] and energy production[280] to boot. With all that in mind, do you remember the Charlie Munger quote from the beginning of this chapter?

"Never, ever, think about something else when you should be thinking about the power of incentives."

When we focus on greed, or motivations, or outcomes, we are "thinking about something else," as Mr. Munger says. Now recall from the section on Money, fiat currency (paper debt) is created when a bank loans money. This is roughly true, whether the bank is lending to itself, to the government, to a business, or to you. Then also recall from the section on Power, the name of the game we are all playing is tournament-style, competitive resource capture. In the very beginning, the resource to capture was other people's stuff—fruit, wheat, shoes, goats—what have you. As that capture game matured, the resource became land and other people's stuff. Further refinements to the game made the primary resources to capture gold and land. Today, the latest version of the game makes the resource sovereign issued paper debt. With enough paper debt, you can easily acquire all of the other pieces on the game board.

If you trace that all out, the game is to capture as much sovereign paper debt as you can. And, the way the rules have evolved, there is one primary source where powerful players can compete to acquire nearly unlimited sovereign paper debt: sovereign controlled banks. If the 2008 global financial collapse taught us anything, it was that if you become leviathan enough,[281] the sovereign will do almost anything to keep you afloat.[282] If we think about the power of incentives, what we can see from the "bailouts" of the last 40 years, the real incentive for all major players in the game is to become systemically important. To quote Mr. Charlie Munger again,[283] "Show me the incentive and I'll show you the outcome." And, if we work backwards from that with Elon Musk, the outcomes lead me to believe he is following the exact same incentives that every other major corporate behemoth is: mass centralization of the sovereign's critical resources.[284] Or, put another way, he appears to be doing exactly what all of the other megacorporations are doing. They are all aligning themselves in such a way that they can essentially hold the sovereign hostage. Why do this?

Because that is the prize. To become the sovereign. To loosely quote Ayn Rand again,[285] "Whoever says you must sacrifice to the state either is, or wants to be, the state." Which traces us back to the discussion on power. The sovereign became sovereign because, through tournament-style competition, they wield a monopoly on coercive force within a geographically bounded area. Today, owing in large part to nuclear weapons, maintaining control of geographically bounded areas is a little trickier.²⁸⁶ Realistically speaking, there are only three truly geographically sovereign nations left:

1. The United States
2. China
3. Russia

No non-nuclear armed military can impose territorial control over any of those countries by force. Moreover, any nuclear armed military force attempting to do so faces the risk of mutual assured destruction.²⁸⁷ So, if you want to impose control of those territories, the only real avenues left are internal—either through politics, economics, or both. Remember, the name of the game is competitive resource capture. The way the rules of the game have developed over the centuries has created a very clear path to capture sovereign power: Gaining unlimited, unfettered access to sovereign issued paper debt.

Or, in the colloquial understanding, positioning yourself at the head of the line to pull fresh sheets off of the "money printer." Which is, ultimately, a full-circle return to the grassroots of the game. A game that started in chaos and evolved into looting, then extortion, and on to the formalized version of extortion we call taxation. The same game that has refined into what is, for most people, a life-long scramble up a greased pole. The same game where the winners always sit atop the pole, set the rules for climbing, and everyone else pays them for the privilege of playing. The incentive today is to do everything you can to make sure the pole collapses if you are not on it. That incentive only exists for one reason, and one reason alone: Because the prize has gradually morphed into sovereign issued debt.

CHAPTER ELEVEN: CHOICES

We can scarcely talk about incentives without spending a little time on choices. Unfortunately, when it comes to questions of choice, this necessitates an inquiry into the realms of rather nebulous topics around things like consciousness and free will.[288] For those that have not bothered to run headlong down that rabbit-hole, the crux of the debate orients around the idea that, if you (one) does not truly have free will, then you cannot truly make a choice. The common perception is, that humans, of course, have free will. Our entire legal system is predicated on the idea that people can and do make choices of their own free will and volition. When we are angry, or even worse, disgusted with someone, it is often because we think and believe that they have chosen to do something offensive to us or someone else. But in that same vein, we also have much capacity for forgiving mistakes.

We also have much capacity for forgiving offense because of the other person's intentions. To give a rough illustration, imagine you see a snake next to your best friend's leg. In an effort to save your friend from the snake, you inadvertently knock your friend off balance. Because of this your friend falls into a puddle, ruining her favorite trousers. However, when your friend falls, it looks comical, so you begin laughing. This hurts your friend's feelings, so she becomes cross with you. When you see she is cross with you, you protest. You say, "No, I am sorry for laughing, but I saw a snake right next to your leg." Unfortunately, when your friend looks, she finds not a snake, but rather a stick. Your friend, who is still cross, says, "You moron, it is not a snake, it is only a stick. You have ruined my favorite trousers for nothing!" You protest again, saying, "I was not trying to ruin your trousers, I really thought there was a snake." The question I pose is, did you make a choice when you shoved your friend?

Some might say you just reacted and they would not be wrong. Others, like your friend, might say otherwise. She might say you chose to push her, even though you knew there was a puddle, and there was a very low likelihood of the object being a snake. She would not be wrong either—assuming you have free will, that is. If we presume you have free will, would it not follow that you chose to ignore the circumstances and chose an action that was unreasonable under those circumstances? Perhaps. But what if I told you that you are biologically wired to see snakes where there are none? What if I told you that human beings will almost instinctively react to the presence of a snake and that your actions were out of your conscious control? If that is true, then does it not also follow that, at least *sometimes*, you do not have free will? And if that is true, then how do you know when you do and when you do not have free will? Do you lose your free will *only* under circumstances like the snake? Or, can you lose your free will under other circumstances?

What if you are on drugs? What if you are sick? What if you have been misled? What if you are being coerced? What if the threat driving the coercion is fake or unrealistic? What if you are hungry?[289] What if your hormonal levels[290] have changed? What biological processes are involved when you make a choice? Dr. Robert Sapolsky rather elegantly lays this process out in a wonderful lecture. To quote[291] the good doctor:

> *If we're going to make any sense of these complex, context dependent behaviors of ours, we have to look at many layers.*

From there, Dr. Sapolsky goes on to say that, if we are truly going to understand human behavior and choice, we must look one second before the decision, seconds-to-minutes before, hours and days before, through childhood and, ultimately, back millions of years. By the end of the lecture, and in response to an audience question, Dr. Sapolsky says:[292]

> *I don't think it's possible to look at this whole range of ways in which our behavior is being shaped by biology—I don't*

see there being a shred of possibility of free will being in there.

He goes on to say he has no problem with the notion of humans not having free will and no idea how the world can possibly work that way. I agree. It is highly problematic to consider a global society where people are not actually making choices rooted in their own volitional free will. Leaving that aside for the moment, this all begs the question, "What does this have to do with greed?" In a word, I would say, "Everything."

Greed is context dependent. Incentives are context dependent. Choices are context dependent. And, within those spheres, all of the context-dependent decisions that are made occur in layer, upon layer, upon layer. That said, in the greater discussion around Power, Money, Greed and Bitcoin, it is important that we evaluate as many layers as we can, as objectively as we are able. This, in a nut-shell, is the entire purpose of the first three parts of this book. It is a quest to better, and more objectively, understand what happened a second before, what happened an hour before, what happened a millennia before the game began. Indeed, to truly be thoughtful on the matter, we should explore what happened 250,000 years before. I think this is important, because we all live on this planet. As such, and the way things are set up, we are all subject to roughly the same rules of roughly the same game that we are all simultaneously and continuously playing—the game of competitive resource capture and control. In that game, we have billions of context-dependent, layered decisions made by billions of people. Among those billions of people, a relatively small number make the most consequential decisions for everyone else. And they are just as bad at making those decisions as you and I are.

They are subject to the same biological forces, the same incentive structures, the same biases, the same emotions, the same complexity—the same everything. While I was at the London School of Economics, another Masters candidate lived above my flat. His research was on conspiracy theories and why they hold such enduring appeal. His conclusion was it is more comfortable for people to imagine being subjected to intentionally evil plans than it is for people to exist knowing all the horrible things that happen in the world come from chaos. Put another way, he reasoned it is more comforting for people to think an evil person is in control, rather than knowing everything is completely out of anyone's control. When it comes to perceptions or accusations of greed, I think a similar thought process occurs.

Take, for example, the United States Federal Reserve (the Fed). In author G. Edward Griffin's book, *The Creature from Jekyll Island*,²⁹³ Griffin contends the Federal Reserve was created from a "Master plan which was designed from top to bottom to serve private interests at the expense of the public." The gist of the book revolves around the idea that the Federal Reserve was not created in response to banking crises and unstable banking, as is commonly portrayed in the economics textbooks. The secret meeting at Jekyll Island is, instead, portrayed as a conspiracy by a tiny handful of powerful men to establish a banking cartel. To do this, they drew up a "master plan" that gave them an unlimited and hidden ability to siphon wealth from John Q. Public. As noted in the section on Money, it is hard to say that the Federal Reserve system has created anything but an absolute mess of things. Yet, Mr. Griffin's contention also violates Hanlon's Razor—a razor that I am quite fond of.

Of course, no one alive today was at the meeting on Jekyll Island. Likewise, no one alive today participated in the founding and creation of the Federal Reserve. Nor is anyone alive that took part in the associated legal and political processes that occurred at that time. What we do have— among volumes of literature about this topic from that period—is a contemporaneous writing made shortly after the creation of the Federal Reserve. Henry Parker Willis wrote *The Federal Reserve System*²⁹⁴ in 1923 —ten years after the creation of the Federal Reserve. To his credit, Mr. Willis does a fine job of laying out the political scene from the perspective of a person on the ground in the years leading up to, and after, the formation of the Fed. The tale Mr. Willis weaves, by contrast, is not one of an elite group secreting off to a mysterious island to lay out plans for a banking cartel. Rather, Willis paints a portrait of a gridlocked Congress, unwilling to tackle the numerous issues surrounding the issuance and redemption of so-called "greenbacks"²⁹⁵ during and after the Civil War, completely misaligned banking interests between state and federally chartered institutions and struggles over the setting of value. To wit, Willis writes:²⁹⁶

> *The growth of a highly individualized system of banking management was in these circumstances to be expected. In later years, when banking reform discussion had reached a more advanced stage, the reluctance, or even refusal, of the more influential bankers of the country to accept any responsibility for institutions other than their own was an outstanding element in the general situation.*
>
> *In the beginning of the banking reform discussion, this reluctance was still nebulous; and, had there been an effort to direct banking thought into scientific channels, a very much earlier advance toward actual improvement might have been made. The struggle over the question of a standard of value, the attempt to substitute the silver dollar—worth at the time perhaps 50 per cent of the gold dollar, which had in effect come to be the established standard of value—is of political rather than of economic, or banking, significance.*

In fact, Willis was the first Secretary of the Federal Reserve.[297] Fans of *Jekyll Island* will, no doubt, assume this makes Mr. Willis an unreliable source. I, on the other hand, find his writing to be thorough and thoughtful. It also happens to align well with very solid modern scholarship on American banking. Calomiris and Haber[298] paint a very similar picture of a deeply fractured unit-banking[299] system that was torn between sharply divided factions and, as a result, was also incredibly fragile. As noted in the section on Money, the United States banking system was formed from rough and tumble beginnings. Keep in mind, the country had just come out of a complete, all-out shooting Civil War only 40 years before the founding of the Fed. For a bit of perspective, Abraham Lincoln was as distant a memory to the founders of the Fed as Ronald Reagan would be for you and me. Nevertheless, since the founding of the country, and very much unlike the "old world," populist agrarian banking interests were powerful in the United States. They wielded as much—if not more—political capital and sway as the so-called "elites" of the old world banking order Griffin wrote about in *Jekyll Island*. Republican politicians of the time ran on a platform of sound money. However, when it came time to actually implement those sound money fixes, they routinely punted.[300] Why? Because those same, bitterly divided urban and agrarian banking elites did not want the

Republicans (or anyone else) tinkering with their profit models—economic stability be damned. With that political and economic gridlock in mind, Calomiris and Haber[301] go on to say:

> *The creation of the Fed required a set of compromises that reflected the power of particular constituencies rather than the economic intentions of the Fed founders. First, despite the recognition of the superiority of a branch-banking system, the founding of the Fed did nothing to reform the particular market structure of America's unit-banking system. That was not an unintended consequence: the creation of the Fed required the support of unit bankers and their political allies.*

Let us pause here for a moment and trace back to the conversation about Elon Musk in the last chapter. Elon Musk has hidden information: his motivations for acquiring massive wealth and power. A conspiratorial person could very easily create a narrative much like the one I crafted in the previous chapter. The one where Elon Musk is secretly a diabolical genius, hell-bent on world domination so he can hold us all hostage for $1 million. In fact, a clever and well-versed conspiratorial person could even write a book about it. They could pull in all kinds of anecdotes, interviews with friends, emails from the past, place it all in any context they desire, and easily paint a *very reasonable* picture of an Evil Dr. Elon Musk. Likewise, another biographer could also paint a picture of Elon Musk that shows a man deeply committed to saving humanity, valiantly struggling against legacy car manufacturers and the legal apparatus that supports them while locking upstart competitors like Musk out. This biographer could show a man that overcame political obstacles, financial obstacles, mockery, disbelief, and near bankruptcy to create an industrial empire. A *very reasonable* narrative that paints this savior of humanity as a man who lives frugally and humbly, all while his eyes are firmly affixed on the long-term goal of making mankind multiplanetary. I posit the same possibilities are here for the Federal Reserve.

A conspiratorial person can easily write a book that sounds quite reasonable. As the story goes, a group of powerful, elite men met in secret to implement a plan for the domination of the entire global financial system, while ensuring their legacy wealth remains unchallenged.

Likewise, another person could paint a picture of a group of people facing a deeply divided country, healing from a catastrophic civil war, struggling with a fractured, unstable banking system and a political class hell-bent on protecting their cronies. A group of men trying desperately to find a way to patch up the bloody mess, so they can stabilize the national finances, and finally bring the USA into the modern era of science. Keeping in mind, of course, these men are also trying to do all of this right on the heels of the massively debt-inducing Spanish-American War. Would it not be *possible* that the Federal Reserve really was a compromise solution that just barely got over the line politically? Would it not be *possible* that powerful men might meet in secret to try and figure out a way to make that work because they could all recognize the system they had was a mess? There is that hidden information problem we were talking about.

The point is, it is quite easy to manufacture a narrative that grants powerful people supernatural abilities to effortlessly guide world events. The reality is often quite a bit less fantastic. The reality is political and economic systems in competitive resource capture economies are highly complex, deeply interwoven, and have incredible momentum driving them. In the midst of that complex, interwoven mess are very fallible, imperfect, biased, moronic, short-sighted, overconfident people—just like you and me. People doing their level best to make decisions and implement solutions that are almost always terrible, never complete, and often leave more problems in their wake than they solve. Of course, this does not mean that every person in that mix is doing everything with the best intentions. It does mean that the people trying to do the "right" thing also have a problem with hidden information. In their case, the hidden information crops up when they are dealing with every other person involved in the endeavor that may or may not be trying to do the "right" thing as well. Remember Hanlon: "Never attribute to malice that which can be adequately explained by incompetence or stupidity."

When we talk about the choices people make, whether at the Federal Reserve, in the halls of Congress, or in the board rooms of the major corporations, it is imperative we judge objectively. But that objectivity requires humility on our part. The recognition that we do not have all the information. The understanding that, more often than not, there are no good choices to make. The allowance for the unanticipated or unintended knock-on effects for every choice that does get made. And, ultimately, the awareness that these choices, these decisions and these outcomes are wholly incentivized by the game we are all playing.

The flawed incentives are inherent to the design of the game. As alluded to in the section on Power, human beings managed to live quite well for 250,000 years without engaging in coordinated warfare. They managed to exist without market economies. They managed to exist without money. There are still tribes of people[302] today that live much the same way. Nevertheless, about 12,000 years ago is arguably when the trouble begins. That trouble just happens to coincide with the time that agriculture and animal husbandry are introduced into human society and, ultimately, is spread out among hunter-gatherers.[303] And, depending on who you ask,[304] the development of these practices could invoke ancient, unknown societies sharing knowledge, a gradual understanding of farming, or some other means of skill acquisition. Regardless of the origins, what is generally clear from the research is agricultural practices and animal husbandry resulted in far higher population densities, which consisted of far unhealthier humans.[305] Meaning, despite agricultural practices being suboptimal for individual human health, they are supremely optimized for broad scope human thriving. With all of this in mind, I think a few things that have been asserted thus far can be even more comfortably reasserted here:

1. Agriculture and animal husbandry became widespread
2. Which created conditions primed for violent conflict
3. That resulted in the creation of rules to manage those conflicts
4. And also created conditions primed for human thriving
5. Which resulted in increasingly large and complex rule-based societies.

What we might deduce from this are a couple of things. Given the scant, but compelling evidence that widespread, coordinated conflict was low to non-existent among hunter-gatherers, it is possible that either agriculture creates a resource capture incentive, or agriculture creates so many people that they inevitably just start fighting. From my perspective, and given that we have thousands of modern examples of incredible concentrations of people living side-by-side in peace, I do not think it is simply because the population grew. Moreover, as I found in one of my early research papers[306] on intentional homicide, the highest rates of homicidal violence occurred in poor countries with high income-inequality and high rates of adherence to vengeance based Abrahamic religious traditions. Put another way, what that paper, and indeed what history has shown is, if you want to create conditions for violent conflict, make most people poor, have a few at the top with a lot, and have a deep admiration for the notion of taking an eye for an eye.

The trouble is, those conditions are apparently only created in competitive resource capture economies. The even larger trouble is, because it has been a game of competitive resource capture for so long, it is taken for granted that this is all "normal." But that is an erroneous assumption. The entire structure of modern society may well be built upon what essentially amounts to a several thousand year process of compounding errors. Just going by pure health metrics,[307] hunter-gatherers work less, are better adjusted emotionally, and are more physically fit and healthy than their agriculturally based counterparts. Hunter-gatherers were also a very stable population—estimated to be around 6–10 million globally[308]—for about 150,000 years. There was a population "boom" some 50,000 years ago, as tool making and clothing improved. But the human population absolutely exploded with agriculture 12,000 years ago. This is also when we started competing for resources. It rather appears that competition for resources in an agricultural setting is an abnormality from a behavioral standpoint. The remarkable thing about it all is, despite the historical abnormality, the rules for competition have now been so thoroughly refined, we have collectively managed to stumble right up to the gates of making the game work.

Decisions, decisions

Rationality is a widely utilized assumption in economics. Rationality also just happens to be an arguably less than useful, and grossly over-utilized[309] term in economics. As Professor Hammond says in the linked paper, "Constructing a realistic descriptive model of behavior is perhaps more of a task for psychologists than for economists." Yet, armed with this admission, Hammond and indeed, the entire field of economics, goes bounding off into the woods in search of said model. Once immeshed in the dark wilds of the human psyche, they desperately search for the rhymes and reasons behind why people make economic decisions. From the Econ 101 normative foundation[310] of rational actors maximizing outcomes, the theories have morphed to ideas of bounded rationality, on to game theoretic Nash Equilibriums, various stage-based models, before middling through the minefield of heuristics and biases, and then trundled off into subjective expected utilities and other "descriptive" models. All the while, and underneath it all, is the rather uncomfortable proposition that people just do stuff because they think it is a good idea at the time, with the even less flattering prospect that what people think is often quite dumb.

Ultimately, understanding the process of making a decision is arguably far less impactful than the results of the decisions that are made. Nevertheless, the biggest problem with all of the various and varied economic theories is the niggling little problem of falsifiability.[311] For those unfamiliar, the idea revolves around the criterion for scientific inquiry put forth by Sir Karl Popper[312] that says a theory can only truly be scientific if it is possible to prove false. It would be a near impossibility to, for instance, falsify the concept of "subjective expected utility." It would be impossible because there would always be an observable instance of someone behaving according to that criteria and no objective way to disprove the cause of that behavior. Put in a more disparaging light, it seems to me the entire field of economics more appropriately belongs in the realm of the metaphysical inquiry[313] rather than any scientific one.

What is readily observable, however, is the ability for an actor to be able to reliably, and repeatedly, modify the behavior of another actor by altering their incentives. "Give me one of your pigs or I will bash your head in" is a brutish, but simple example of incentive modification. Of course, one could fritter about, extolling the process of this economic actor determining the "subjective expected utility" of not getting conked on the head versus giving up a pig. The observable reality is there is a high likelihood of compliance with that demand, which only increases with the credibility of the threat. Moreover, there is also an observable reality that says a select few will resist, perhaps to their serious detriment—and, then again, perhaps not. Indeed, it is quite possible they will prevail. The broad point being, much of the behavior explored by the economists and philosophers is likely closer to gambling than anything else.

As a gambler finds in a casino, the house has a number of incentives they can use to encourage or retard the behavior of their patrons. Indeed, many of those incentive structures have been well documented and researched.[314] An infamous and related example even ended up as the subject[315] of a Nevada Gaming Commission court case. The core structure of the casino environment is one that is constructed entirely with gambler incentives in mind. Everything from the decor,[316] to the food, to the floor design and lighting and everything in-between—they are all crafted to incentivize gamblers to stay as long as possible. Even when they are broke, or tired, or indeed, spending their next month's rent. Keeping in mind, of course, that this environment is built within the confines of enormous structures that clearly cost a lot of money to build and maintain. Meaning, everyone who walks through the door is, or should be, immediately on notice that the profit margins are clearly high enough to warrant these exorbitant expenses and that the money in their pocket is the only possible source of those profits.

Within the discussion at hand, when we are talking about greed in the human structures of power and money, there is a distinct line of scholarship devoted to the idea that the entire pseudo-capitalist structure underpinning the global political economy is really nothing more than a giant casino.[317] Unlike a walk-in casino, however, the global political economy casino is not one we can simply leave. Rather, the opposite is true, and the sovereign—your sovereign—demands that we all both play and pay under penalty of death. As Susan Strange writes in *Casino Capitalism*:[318]

> *The Western financial system is rapidly coming to resemble nothing as much as a vast casino. Every day games are played in this casino that involve sums of money so large that they cannot be imagined. At night the games go on at the other side of the world. In the towering office blocks that dominate all the great cities of the world, rooms are full of chain-smoking young men all playing these games. Their eyes are fixed on computer screens flickering with changing prices. They play by intercontinental telephone or by tapping electronic machines. They are just like the gamblers in casinos watching the clicking spin of a silver ball on a roulette wheel.*

It is, perhaps, disingenuous to relegate the entirety of the global political economy to the ranks of a vast 24-hour casino. But the underlying logic may certainly help explain some of the madness that is otherwise attributed to nefarious secret cabals of lizard people. The reason this underlying logic may prove helpful relates to a notion many find quite troubling. The notion that much of what passes for success may, in fact, be more attributable to simply getting lucky.

CHAPTER TWELVE: CHAOS AND LUCK

There is an old saying, "Luck is what happens when preparation meets opportunity." It is widely credited to Seneca,[319] but the provenance of the quote is disputed.[320] A similar quote, attributed to Samuel Goldwyn[321] says, "The harder I work, the luckier I get." In the first quote, the core idea is, if you are sufficiently prepared, when an opportunity comes along, you will be in a position to capitalize upon it. In the second quote, the core idea is, if you work hard enough, more opportunities will present themselves. These are *perfectly reasonable* assumptions. They also comport well with common experience. Intuitively, they also serve as a mild cautionary against another deadly sin, sloth.[322]

Obviously, if you are indolently lazing about, the likelihood of you becoming successful is dramatically decreased. This, of course, depends heavily on your definition of success. Nevertheless, begging for beer money on a street corner is clearly not going to prime you for successful outcomes. That is one end of the scale. At the other end of the scale, however, we also run into problems. Is it possible to work incredibly hard and still be grotesquely unlucky? Of course. It is indisputable. Likewise, is it possible to be incredibly well prepared for life and never have an opportunity to put that preparation to good use? Certainly. This too is indisputable. The point being, I think it largely indisputable that luck plays some role in everyone's lives, good or bad.

Yet when we are talking about luck, what exactly are we talking about? Are we considering luck a force, a circumstance, or plain randomness? Likely, we are discussing all three at once, depending on the context, our particular ideological or philosophical bend, and how superstitious (or not) we may be. And if this is true, then perhaps, the most troubling aspect of luck is that it is impossible to pin down. The thing is, the core concepts are not locked into isolation when discussing human interaction. Indeed, chaos theory[323] is very much rooted in very similar questions. Is chaos a force, a circumstance, or plain randomness? The reason I invoke luck here is because of the little problem we humans have with causal inferences[324] and patternicity.[325] Just as an aside, patternicity is also referred to in the medical literature as apophenia.[326]

Causality and causation are slightly different sides of the causal coin. Causality[327] is "an existing relationship between the effect and what it was caused by (object, state, or process)." By contrast, causation simply means to be the cause of something. In legal terms, causation is further broken down into factual, or actual causation and proximate causation. As a first-year law student can tell you, the "but-for"[328] test is the means by which a court will determine actual cause. Meaning, but for the act, the result would not have occurred. While this may sound simple enough, legally determining cause is actually a bit of a trip down the old rabbit-hole.[329] This is where the concept of proximate causation comes in. Proximate, or legal cause, tries to roughly sketch out the boundaries of causation from a legal liability standpoint.

For example, you could say, "But for John Wilkes Booth[330] being born, Abraham Lincoln would not have been shot." If we were really pedantic about things, we could say that John Wilkes Booth's parents were the "cause" of Lincoln's assassination. Of course, no one in their right mind would suggest this is a fair, or even logical outcome. This is where proximate or legal cause comes in. Within that line of reasoning, there must be an identifiable sequential connection between the cause and the outcome. For instance, in the John Wilkes Booth scenario, the sequence of events between his birth and his decision to shoot Lincoln is very far removed. Meaning, that time span breaks the causal chain of events. Nevertheless, in this example, the cause is relatively straight-forward. But for John Wilkes Booth pulling the trigger, Lincoln would not have been shot. Easy enough. But what about this scenario? The following is taken from a famous tort law case, *Palsgraf v. Long Island Rail Road*:[331]

> *Plaintiff was standing on a platform of defendant's railroad after buying a ticket to go to Rockaway Beach. A train stopped at the station, bound for another place. Two men ran forward to catch it. One of the men reached the platform of the car without mishap, though the train was already moving. The other man, carrying a package, jumped aboard the car, but seemed unsteady as if about to fall. A guard on the car, who had held the door open, reached forward to help him in, and another guard on the platform pushed him from behind.*
>
> *In this act, the package was dislodged, and fell upon the rails. It was a package of small size, about fifteen inches long, and was covered by a newspaper. In fact it contained fireworks, but there was nothing in its appearance to give notice of its contents. The fireworks when they fell exploded. The shock of the explosion threw down some scales at the other end of the platform, many feet away. The scales struck the plaintiff, causing injuries for which she sues.*

The question is, what caused her injury? The scales falling, obviously. But what caused the scales to fall? The fireworks, right? But what caused the fireworks to go off? They went off when they hit the ground. So, gravity is the culprit there. But what caused the fireworks to hit the ground? Was it the pull from within by the guard, or the push from without by the other guard? Was it the man being late and trying to catch the train? Was it because he did not pack his fireworks very well? It gets a little tricky, does it not? This is important in context, because it is generally within the realms of scientific inquiry or in the assignment of legal liability that causal claims are ever rigorously tested. This is arguably why so few truly understand causal mechanisms. Or, rather it should be said, so few are ever confronted with a need sufficient enough to objectively map out a causal connection.

Meaning, when it comes to determining causality, overly simplistic heuristics[332] tend to dominate. Nevertheless, and returning to chaos theory, a popular idea emerged a few decades ago, colloquially referred to as the

Butterfly Effect.³³³ The pop-culture understanding of the Butterfly Effect emphasizes large outcomes from minute inputs. This is evidenced by a movie reference in *Havana*, where the star says,³³⁴ "a butterfly can flutter its wings over a flower in China and cause a hurricane in the Caribbean." Indeed, this version of the idea shows up in a number of films, including a film named *The Butterfly Effect*.³³⁵ But, as the linked article notes, the author's intent was rather to:

> *Illustrate the idea that some complex dynamical systems exhibit unpredictable behaviors such that small variances in the initial conditions could have profound and widely divergent effects on the system's outcomes.*

I bring this all to your attention here because causality, causation, and indeed, luck all have a profound impact on context and perception. This is especially true in the realm of success, which becomes even more important to understand in the context of a resource capture game predicated on violence. Recall from earlier chapters, there is a common perception and understanding of the world that says, "Life is a competition for scarce resources."³³⁶ This idea has a certain intuitive appeal and it ignores one very real possibility that the entire earth may well fit the definition of a singular organism. And if that is true, then this "competition" may actually be a means to achieve and maintain homeostasis within that organism. Again, if we are to invoke some sort of "first principles," indeed, it would be imperative to sort this issue out.

Since no one has—or is even able to—then we are left with subjective observations and academic dithering and debate to guide any inquiry. I return to this idea here because there is a natural tendency to assign some form, or combination, of causal factors to successful participants in any competition. Whether it is football, poker, physical combat or business, if you ask the winners, "Why were you successful today?" they will invariably point to something they did to bring about the outcome. They will say they worked hard, trained harder, were more focused, more dedicated, had God on their side—the list is virtually endless. What you will rarely hear is someone say, "I just got lucky." Yet given the endless possible combinations of factors in any competition, this is almost always the most honest answer. I give Jeff Bezos an enormous amount of credit for admitting as much[337] about the success of Amazon—a position he maintains to this day.

On a related note, there is a young man that rose to internet infamy by declaring he was a millionaire and was going to chuck it all to start over. His intention was to demonstrate how he could make a million dollars again starting from nothing. He even dubbed it the "Million Dollar Comeback." Spoiler alert, the experiment did not quite work out that way.[338] Rather, after enduring enormous hardship, he made about $65,000 and gave up. Keep in mind, this young man, by all outward observations, has all of the essential character traits[339] that are generally embodied by successful people. He appears to be very open, conscientious, agreeable, extroverted and low in neuroticism. He is also a young man that has entrepreneurial experience, is technologically capable, highly intelligent and is young and in good health.

Yet despite all of this, he catastrophically failed at achieving his stated goals. While people may, and do, criticize his intentions, his ability, the veracity of the experiment and any number of other things, his core problem can be boiled down to survivorship bias.[340] One of the more famous examples of survivorship bias occurred during World War II. Allied air commanders evaluated planes that had been shot up and returned to the airfield. They observed where the majority of bullet and fragmentation holes were located and then decided to add armor to those sections. A mathematician named Abraham Wald pointed out their error.[341] They were selecting areas to add armor based on planes that made it back. What they needed to do was add armor to the areas of the plane that were hit and caused the plane to go down. Meaning, their bias in determining armor placement occurred because they were incorrectly relying upon the survivors of a selection process and failed to account for the failures.

Survivorship bias shows up everywhere. It occurs in finance, sales, mate selection, you name it. The net effect is people misattribute a causal relationship based on a highly-biased sample—namely the survivors of a selection process. To put this into the context of the "Comeback Millionaire," the mistake he made was to assume his success was solely related to his inputs, namely hard work, dedication, and determination. What he failed to recognize are the thousands and thousands of other similarly situated, similarly possessed people that undertake the same endeavors and fail. Meaning, he survived a selection process (becoming a millionaire at a young age) and assumed that is replicable because he failed to account for all the people that have tried and failed to do the same.

The same process can roughly be observed in reverse. The "Swiss Cheese Model"[342] describes a series of security failures that result in a harm occurring. This is commonly invoked in medical settings, but is also applicable to industrial accidents, airplane crashes, and indeed, even the fight sciences. The idea is, safety and security processes are implemented in layers. The reason for this is so that one failure, or even multiple failures, in safety and security procedures does not result in a catastrophic system failure. In the case of an air crash, for instance, planes do not simply fall out of the sky for no reason. Rather, a crash is often precipitated by multiple failures on multiple levels. With the right combination, the holes in the Swiss cheese can eventually line up and create conditions where a crash becomes inevitable.

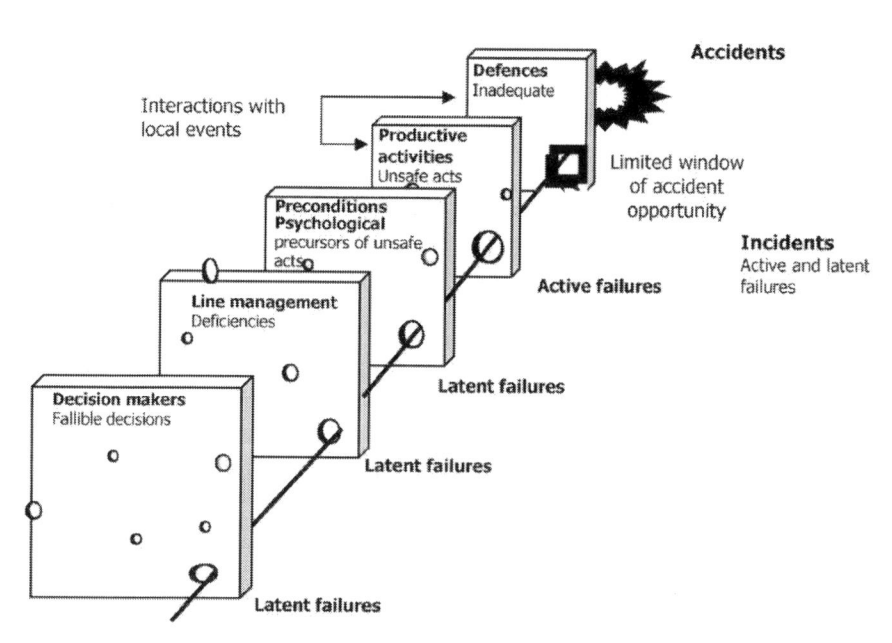

This is the rough reverse of survivorship bias because it implicates a very similar process for success. Meaning, to be successful in any competitive environment, on some level, the holes in the cheese have to line up in your favor in order for you to prevail in the competition. Survivorship bias shows up when people assume the holes will always line up in their favor because the holes lined up for them in the past. When we are talking about a game of competitive resource capture, however, survivorship bias becomes highly problematic. This is especially true when the competition itself is predicated on the use of coercive violence to capture those resources.

In any sort of combat, the most trivial of input alterations can have outsize impacts on combat outcomes. For example, in a well-publicized mixed martial arts (MMA) match between Jose Aldo and Conor McGregor, the two combatants entered the ring as evenly matched as can practically be accomplished. Both men spent an enormous amount of time and resources to prepare for the match. Both men had a demonstrated track record of success in competitive MMA. And when the bell rang, Mr. McGregor knocked Mr. Aldo out with one punch[343] 13-seconds into the match. Keep in mind, this is a highly controlled environment, where the competitors are literally weighed before the competition in an effort to ensure the fairest possible fight. Had Mr. Aldo's head been turned slightly towards one direction or another, the fight may well have continued. Had Mr. McGregor thrown the punch a quarter-second later, perhaps the fight would have continued as well. This is not even mentioning that Mr. Aldo also connected with a very similar punch milliseconds after Mr. McGregor's punch landed. On that night, in that ring, the holes in the Swiss cheese lined up against Mr. Aldo. Or, conversely, the holes in the Swiss cheese of success lined up in favor of Mr. McGregor. It really just depends on your point of view.

The point being, leading up to the outcome, there were unlimited possibilities that could never be fully accounted for by any amount of training, preparation, dedication or determination. There is an old adage among gunfighters that says, "Gunfighting is a game where you can do everything right and still lose." This adage holds true in every single instance of violent combat because forces of chaos can, and routinely do, overwhelm the best of preparations. In the context of the discussion at hand, the trouble arises when the victors of violent resource capture competitions assume, or presume, their success is attributable to their innate ability, divine providence, purity of purpose, or any number of other things that are far better explained by forces of chaos. Meaning, there is no concrete, legitimate reason to be deferential to the victors of any competition. They may perform admirable feats. They may be exceptionally dedicated or determined. But that does not inherently distinguish them from the thousands, if not hundreds of thousands of other people that have done the same, but were not victorious, or even particularly successful.

Yet our society, and indeed, the entire global political economy is predicated on exactly that model. A model that reifies the demonstrably false idea that, "only the strongest survive," or that the "cream always rises to the top."[344] This simply is not true. The more accurate, and certainly less flattering reality is, more often than not, the Swiss cheese just happened to line up for the winners. And once in the winner's circle, no one pays attention to the thousands and thousands of similarly talented, similarly situated losers left in their wake. But just because the Swiss cheese lined up does not necessarily impart some sort of advanced knowledge, ability, skill, or anything else to make those winners any better than any of their competitors to make rules. In fact, and more likely than not, the people deeply embroiled in the trappings of survivorship bias are probably some of the least best people to be making rules for the governance of a highly complex agrarian society.

Much like discussions around free will, any discourse on the role of luck in success is often met with vehement disagreement. I think even entertaining the notion runs afoul of deeply held beliefs about human beings and the fundamental nature of reality. This is perfectly understandable. At smaller scales, the notion of putting forth effort and seeing the resultant output is a readily observable reality. If I take the time to learn how to woodwork and then build a chair, it would be ridiculous to say it was luck that created the chair. Professor Robert H. Frank[345] of Cornell does a wonderful job of breaking down the role of luck in his book, *Success and Luck: Good Fortune and the Myth of Meritocracy*. I mention it here because it does a fine job of highlighting some of the more problematic cognitive biases that lead to the reflexive revulsion against the notion of luck being a determining factor in success.

Nevertheless, I will say, of course, it is not simply a matter of luck for someone in the middle of the curve[346] to achieve success that has a proximate causal relationship to their effort. What I am demonstrating here relates to the tails of the curve—thus, the Swiss cheese model. The takeaway should be that, when we are talking about very rare success in very complex competitions, it is a fundamental error to presume it is the conscious inputs of the victorious that led to the outcome. As Frank writes about Bill Gates:[347]

> *In short, most of us would never have heard of Microsoft if any one of a long sequence of improbable events had not occurred. If Bill Gates had been born in 1945 rather than 1955, if his high school had not had a computer lab with one of the terminals that could offer instant feedback, if IBM had reached an agreement with Gary Kildall's Digital Research, or if Tim Patterson had been a more experienced negotiator, Gates almost certainly would not have succeeded on such a grand scale.*

The point being, when it comes to things like outcomes of the great wars in history, political machinations, the outcomes of assassinations and assassination attempts, the squirrelly paths of seizing power, the leaders that emerge and the rules they create—these cannot coherently be attributed to anyone's particular personal characteristics. Under the influence of hindsight bias,[348] they can become the source of ideological and political narratives, but they hold no meaningful place in a rational discourse about economics, political economy, or indeed, money. Which is to further say, and keeping Hanlon's Razor in mind, much of what is attributed to greed is just as likely attributable to survivorship and hindsight biases that favor the victors in competitions that are largely determined by luck. Or, put simply, when we reify the victors, as a practical matter, we may just as well be reifying an exceptional lottery winner.

CHAPTER THIRTEEN: WASTE

In Blackstone's *Commentaries on the Laws of England*,[349] Blackstone briefly and sporadically touches upon the duty of people born into the nobility. Writing first that the "inferior" classes may not have the time or inclination to study the rules, "Those, on whom nature and fortune have bestowed more abilities and greater leisure, cannot be so easily excused. These advantages are given them, not for the benefit of themselves only, but also of the public." Bear in mind, this was written during the rapid ascendancy of the British Empire. Thus it would have been at a time when it seemed to many amongst the nobility that Britain truly was destined to rule the globe. Yet in this quote is a reflection of something thoughtful. The idea that the landed and learned classes were not just there to look pretty and squander wealth. Rather, and because of their station, they owed it to the people to use their time and energy to learn about the world and make life better for the people of England.

By the time the sun did finally start setting on the British Empire, that thoughtful purpose had slowly degraded to the rather crude notion that the nobility was just there to look pretty and provide the commoners with jobs keeping their estate houses running. There is a largely accurate portrayal of this attitude on display in the television series *Downton Abbey*.[350] This is all noteworthy here because of what ought to be the readily observable waste of resources involved in creating and supporting a noble class. To put it in perspective, Highclere Castle, which is fictionalized as Downton Abbey in the television show, sits on a 1000 acre tract,[351] while the entirety of the estate sits on 5000 acres.³⁵² Highclere boasts some 200–300 rooms, encompasses 30,000 square-feet, and features neat things like a hand-carved staircase from solid oak, a mile-long driveway, and is worth north of $200 million. An entire village was relocated³⁵³ when the Second Earl of Carnarvon decided the estate needed more room for gardens and whatnot. To maintain this opulent structure requires roughly $1.5 million per year. All of this to give, at its peak, around 100 locals a rather meager existence as employees of the estate.

Keep in mind, the forefathers of those locals would have lived in the "Area of Outstanding natural beauty" that they lost access to when the Earl moved them out. What the Earl got for the commoner's trouble was 2000 arable acres, and around 1500 acres each of park- and woodland. These nobles put this arable land to good use though. They raise a significant crop of oats,³⁵⁴ which they use to supply "leading racehorse owners and trainers." The estate, through rights of heritable property, has been handed down for generations. Meaning, the current occupants own it simply because they were born. They did, quite literally, nothing to enjoy the advantages and life of leisure that Sir William Blackstone spoke about in *Commentaries*. If one steps back and considers the amount of time and labor that has been devoted to the construction of the various structures that occupy this area of outstanding natural beauty it really ought to be quite astonishing. The current structure of the castle was built in the late 1700s. Meaning, all the lumber, the materials, the glass, the nails, everything was built by hand, without power tools. The material was harvested and moved by animal and human power. All in pursuit of the life of leisure that should, according to Blackstone, give these fine folks the time to think about how to make the world a better place.

Oddly, apparently no one stopped to consider that, perhaps, not wasting all of that time and all of those resources to begin with might be a good idea. In terms of improving the lot of the commoners around you, it would seem self-evident that creating enormous estates and manor houses should probably take a back seat to even the barest philanthropy. But, alas, this was not so. Moreover, Highclere Castle just represents one stately home. Throughout the United Kingdom, to this day, are another 3,700 stately homes,³⁵⁵ 95% of which are hereditary and privately owned. Since we are on the broad topic of greed, this is as good a time as any to ask, did the nobility build all of this opulence and grandeur simply out of greed? Is this behavior evidence of "A selfish and excessive desire for more of something (such as money) than is needed?" Or, is this evidence of what Ayn Rand was talking about when she said, "Since time immemorial and pre-industrial, 'greed' has been the accusation hurled at the rich by the concrete-bound illiterates who were unable to conceive of the source of wealth or of the motivation of those who produce it." Is the English nobility evidence of the hard-work and motivation required to create wealth? Or, was Ms. Rand talking about someone or something else?

Were the villagers upended by the Second Earl of Carnarvon merely a bunch of mud-bound illiterates? How many of those mud-bound illiterates supported and reified the Second Earl of Carnarvon when they either agreed, or were forced, to relocate? Are there not millions, if not billions, of people around the world that still admire and follow the English monarchy to this day?[356] Objectively speaking, why should the English nobility command and control vast real estate fortunes[357] and spend such exorbitant sums[358] of public money on funerals and coronations? Are they just greedy little goblins? Of course, it is entirely possible that generation after generation of English nobility are simply people that just happen to be born obsessed with "A selfish and excessive desire for more of something (such as money) than is needed." And it is also entirely possible that generation after generation of English nobility are people that are born naturally able to meet the requirements to be "the source and motivation to create wealth." Neither strikes me as terribly coherent narratives.

This is especially true given the readily observable and, dare I say, catastrophic waste of resources involved in the creation and maintenance of the intricate human social structure of nobility. With this in mind consider, for a moment, this lengthy passage from Sir William Blackstone's Commentaries:[359]

> *If man were to live in a state of nature, unconnected with other individuals, there would be no occasion for any other laws,* ***than the law of nature, and the law of God****. Neither could any other law possibly exist; for a law always supposes some superior who is to make it; and in a state of nature we*

> *are all equal, without any other superior but him who is the author of our being.*
>
> *But man was formed for society; and, as is demonstrated by the writers on this subject, is neither capable of living alone, nor indeed has the courage to do it. However, as it is impossible for the whole race of mankind to be united in one great society, they must necessarily divide into many; and form separate states, commonwealths, and nations; entirely independent of each other, and yet liable to a mutual intercourse.*

Blackstone mentions the "law of nature" 27 times in *Commentaries on the Laws of England*. Bear in mind, *Commentaries* is one of the great texts of the English Common Law. Abraham Lincoln, well known for his self-directed study[360] and eventual acceptance to the bar, almost certainly read Blackstone,[361] despite it being dated by 70 years when he read it. Underpinning the judicial philosophy and common understanding of the rules of the game was the rather (at the time) uncontroversial idea that everything in the world was governed by the "law of nature" or by God.

But what is this law of nature that Blackstone references? It turns out, if we peel back the pages of history a bit, we find the medieval thinkers, philosophers and scientists of the time had a very different conception of things than we do now. Scarcely a surprising statement, and yet, it matters in the context of the discussion at hand. For the medieval scholar, the central idea[362] was that:

> *Everything in Nature had a place, associated with each place was a norm. Different creatures had different norms to which they were expected to adhere: but to every sort, an ideal pattern of behavior was there. Within the embrace of Nature as a whole, every type of creature had its own 'nature', which was a sort of way of conducting itself, a style of being that was proper to it.*

Not only this, everything in nature had a purpose as well. Deviations from these norms, these places and these purposes was thought to bring about negative outcomes, or was a source of calamity. This line of reasoning gave birth to the idea of the laws of nature. This notion says that all things strive to obey their natural behavior, purpose and place. To be clear, this is not to say that medieval thinkers ascribed purposeful action to things like rocks or plants. It is rather closer, but still somewhat inaccurate, to say that everything had a purpose assigned by God. Within that purpose, all things "strived" to adhere to that Godly purpose. Reason, it was then presumed, was there to give guidance to understand the nature of things—to "discern the truth about our nature."[363]

As I risk trundling off into the wilderness here, I shall return to the matter at hand. The broad point of this lengthy exercise is to draw attention to the mechanisms by which medieval thinkers *very reasonably* created incentive structures that favored class distinctions. These mechanisms are printed right on the pages of Blackstone's *Commentaries* for all the economists, historians and anyone else interested in the political economy to see. Yet scant weight is given to them. I highlight them here for the singular purpose of describing why the nobility probably was not greedy, nor were they necessarily particularly adept at being "the source and motivation to create wealth." While we may scoff at the notion, the broadly accepted idea at the time was that all things had a nature to which they must all strive to obey. If an heir, hundreds of years removed from his competitive resource capturing forbearers, is born into that family, does it strain medieval reasoning to believe that his "nature" is going to be different than that of the commoner who was not? If the laws of nature dictate what purpose all things must be striving for, then certainly one born high will have a different purpose than one born low.

Moreover, once it is established and widely accepted that, "it is the laws of nature that dictates some are nobles, while most will not be," then it makes a lot more sense why nobles must do noble things. The laws of nature command it. Likewise, if the laws of nature are such that the commoner is to be beholden and in servitude to the nobles, then of course, it is the natural order of things for them to do so. This, perhaps, may seem trite. Indeed, the original Star Trek series rather overtly[364] explored very similar ideas. Nevertheless, it does not strain credulity to say that, perhaps, the nobles did not build grand palaces and expansive estates because they simply had a "selfish and excessive desire for more of something (such as money) than is needed." Rather, and quite plausibly, they did these things because they *very reasonably* thought that is exactly what the laws of nature compelled them to do. And the commoners basically went along with it because they *very reasonably* thought that is what their nature compelled them to do as well. Likewise, the acquisition and utilization of exotic and valuable resources, the capture of land, the relocation of villages, the demand for opulence was, and is apparently still to this day, not objectionable because it is not perceived as wasteful. It is still, arguably, perceived as adhering to the laws of nature to this day. Indeed, the fact that nation-states still exist lends a lot of weight to the idea that we are all, collectively, still operating under these "laws of nature" that Blackstone laid out nearly 300 years ago. Recall, Sir Blackstone said:

> *However, as it is **impossible** for the whole race of mankind to be united in one great society, they must necessarily divide into many; and form separate states, commonwealths, and nations; entirely independent of each other, and yet liable to a mutual intercourse.*

With that in mind, it begs the question, "Is this true?" If you (or I) randomly asked a stranger on the street, or indeed, if we ask ourselves, "Is it possible for modern society to function without national borders?" What answer do you think will come forth? Given the current political focus on migration and immigration issues, I think it fair to say that most embroiled in that discussion would agree with Blackstone. Yet Blackstone writes these lines in deference to the "laws of nature" that can, and have been, definitively disproved. No scientist today would say with a straight face that rocks have a "nature" they are striving to fulfill.

This is far more important to consider in the grand scheme of things than may be apparent at first blush. When I wrote earlier that the development of modern society may well be a centuries long process of compounding errors, this is but one example. The trouble is, entertaining a discussion such as this is largely academic. There is, quite literally, nothing to be done, save for taking a bit of wisdom from it. The hard reality is, we must work with the world as it exists, not as it could have been, should have been, or otherwise. As noted in the section on Money, there is enormous human thought, effort, energy and momentum embodied in the political economy we find ourselves in today. To forcefully disrupt that can only be calamitous and will undoubtedly result in avoidable harm and much needless suffering. But an awareness of these types of errors, I think, can be of enormous value. This is especially true today, as we all must collectively navigate our inherited, very human created socio-political structures. However, within the context of the discussion about greed broadly and waste narrowly, there are worthy considerations.

If we trace forward from here to a modern issue, we can, perhaps, put this knowledge to good use. Let us use the financial collapse of 2008[365] and the subsequent "bail-outs" as a test case for that claim. To hear the lay-historians tell the story, the 2008 financial collapse was caused by sub-prime lending[366] and a housing crash. I would direct your attention briefly to the issues with this narrative relating to causation and causality. As noted above, few have the requisite need or desire to delve much deeper than this lay-explanation for the "cause" of the financial collapse. On the other side, those responsible for this collapse have every incentive imaginable to further this narrative and do everything they can to obscure the true causes. By the time the Government Accountability Office issued their 663-page report[367] on the causes, most had moved on to the latest tabloid issue of the day. This is not to say that *no one* cared. It is rather to say, the reasons behind the collapse are opaque, difficult for those not well versed in financial lingo to unpack and understand, and the majority of the US population does not even possess the ability to read[368] the report.

Nevertheless, if we unpack the financial crisis a little deeper than the subprime narrative, what we find is quite a bit more alarming. If you are not inclined to read the 663-page report, Matt Taibbi of *Rolling Stone* fame does a pretty good job[369] of encapsulating the madness that led to the greatest financial collapse since the Great Depression. He wrote a series of articles that are certainly read in a much more disparaging light[370] than the GAO report might have ventured. But the substance of Taibbi's reporting rather accurately captures the essence of the GAO version. Indeed, even the authors of the GAO report saw fit to label Chapter 10 "The Madness," followed by Chapter 11, ignominiously labeled "The Bust." The short of the story is, throughout the Clinton administration, and under the guidance of Ayn Rand protégé and Fed Chair Alan Greenspan, a number of financial regulations were rolled back from the Glass-Steagall[371] era. Passed in the wake of the Great Depression, Glass-Steagall prevented, among other things, commercial banks from acting as investment banks and disallowed banks of any kind from participating in the insurance game. And, as noted in the section on Money, this all coincided with advances in computing, and quantitative modeling. The sum of which resulted in some really neat ideas like Collateralized Debt Obligations (CDO) and Credit Default Swaps (CDS).

They were neat because, not only could Wall Street sell these things as risk-management tools to the very few regulators looking over their shoulders, they also had a quality most endearing to Wall Street: They were enormously profitable. By the time Warren Buffet was labeling these complex derivative "assets" as financial weapons of mass destruction,[372] the die had already been cast. The looming trouble with these instruments was magnified by the quantitative nature of the risk models. When you are a Wall Street Bro and you have got a PhD math boffin saying his model predicts only a 1 in 10,000 chance of a bet going south, that makes for a pretty compelling sales narrative. That the Wall Street Bro had not the slightest understanding of the math, the modeling, the assumptions, or anything else was largely irrelevant to them. Compound that lack of understanding across thousands of Wall Street Bros, exponentially multiply the sums of money involved and you end up in a spot where the entirety of the global financial system is teetering on the brink of collapse. The thing that so few realized then, and still fail to realize today is, at the bottom of that risk mitigation ladder is a fat goose egg. There is no counter-party[373] to the enormous amount of leveraged risk that supports the entire global financial system.

Quite literally no one, in this entire sordid cast of characters, from the prudential regulators, to the politicians, to the CEOs, down to the traders at the desks, understood any of this. To a man and woman, they saw the numbers go up, they blustered about as if they knew what the hell they were talking about, and then they collectively brought the entire world to its knees. I mention this all here in context because it speaks directly to the issues I highlighted above. The people that were ostensibly responsible for ensuring that a calamity of this magnitude did not happen were entirely incapable of doing so. They rose to those positions as a result of a selection process that they all collectively survived. The political right in America trends towards the idea that these men and women are there because they "earned" it. In doing so, they become inadvertent victims of survivorship bias. Then, with the benefit of hindsight bias, both the "leaders" and their right leaning supporters assign unearned powers of observation, stewardship, and sound judgment to the winners of this selection process. Not because these people necessarily possess those qualities. It is rather because of the erroneous belief that the selection process had somehow filtered out the bad and left only the good as the victors. But as noted above, that is an incorrect assumption. The less flattering version, which comports much better with reality is, the people that end up in those positions of decision making power and authority are there largely due to dumb luck.

In turn, this may help explain how a group of people, who are supposedly the "cream of the crop," can somehow drive the global financial bus into a ditch without even the slightest inkling they had lost control. Put another way, I argue here that we are collectively making the same mistake our medieval forebears did with the "laws of nature." There, the erroneous assumption was that the nobility were merely following innate laws that compelled them to act as the nobility should. Modernly, in contrast, the suggestion here is that we are all collectively and erroneously assuming competitive selection processes actually work. Meaning, we have, potentially, swapped the "laws of nature" for a similarly faulty presumption and belief in the "laws of competition." This is further evidenced by the fact that the very same architects of this catastrophic global financial failure were the same ones tasked to fix the bloody mess. Much to the public chagrin, these same buffoons were afforded luxurious payouts and bonuses. As Matt Taibbi writes:[374]

> *The following February, when AIG posted $11.5 billion in annual losses, it announced the resignation of Cassano as head of AIGFP, saying an auditor had found a "material weakness" in the CDS portfolio. But amazingly, the company not only allowed Cassano to keep $34 million in bonuses, it kept him on as a consultant for $1 million a month. In fact, Cassano remained on the payroll and kept collecting his monthly million through the end of September 2008, even after taxpayers had been forced to hand AIG $85 billion to patch up his fuck-ups. When asked in October why the*

company still retained Cassano at his $1 million-a-month rate despite his role in the probable downfall of Western civilization, CEO Martin Sullivan told Congress with a straight face that AIG wanted to "retain the 20-year knowledge that Mr. Cassano had."

In defense of the payouts, bonuses and other shenanigans, and despite "punching a hole in the fabric of the universe," Taibbi goes on to report the following conversation with an AIG spokesperson:

*"We, uh, needed to keep these **highly expert** people in their seats," AIG spokeswoman Christina Pretto says to me in early February.*

"But didn't these 'highly expert people' basically destroy your company?" I ask.

Pretto protests, says this isn't fair.

The point being, there was an enormous amount of waste in the bailouts of 2008, much like there was, and is, an enormous amount of waste in the creation and continued support of the English nobility. Much like the English peasantry did throughout the ages, there were protests and public displays of displeasure. But in the end, the noble classes are still there. Just like the moneyed, asset holding classes are still here. On the flip side of all of this is the left leaning take that all of these people are simply driven by naked greed and nothing more. This is certainly a reasonable assumption and it would be hard to say anything to the contrary. And yet, if we look a little deeper, the machinery that makes our global financial system go is anything but simple. Twitter user @concodanomics[375] has created a wonderful series of graphics that detail a number of Federal Reserve, banking and hedge fund market operations. For just one example, the image below covers "The opening leg of a centrally-cleared relative value (RV) trade":

THE GREAT REALIGNMENT

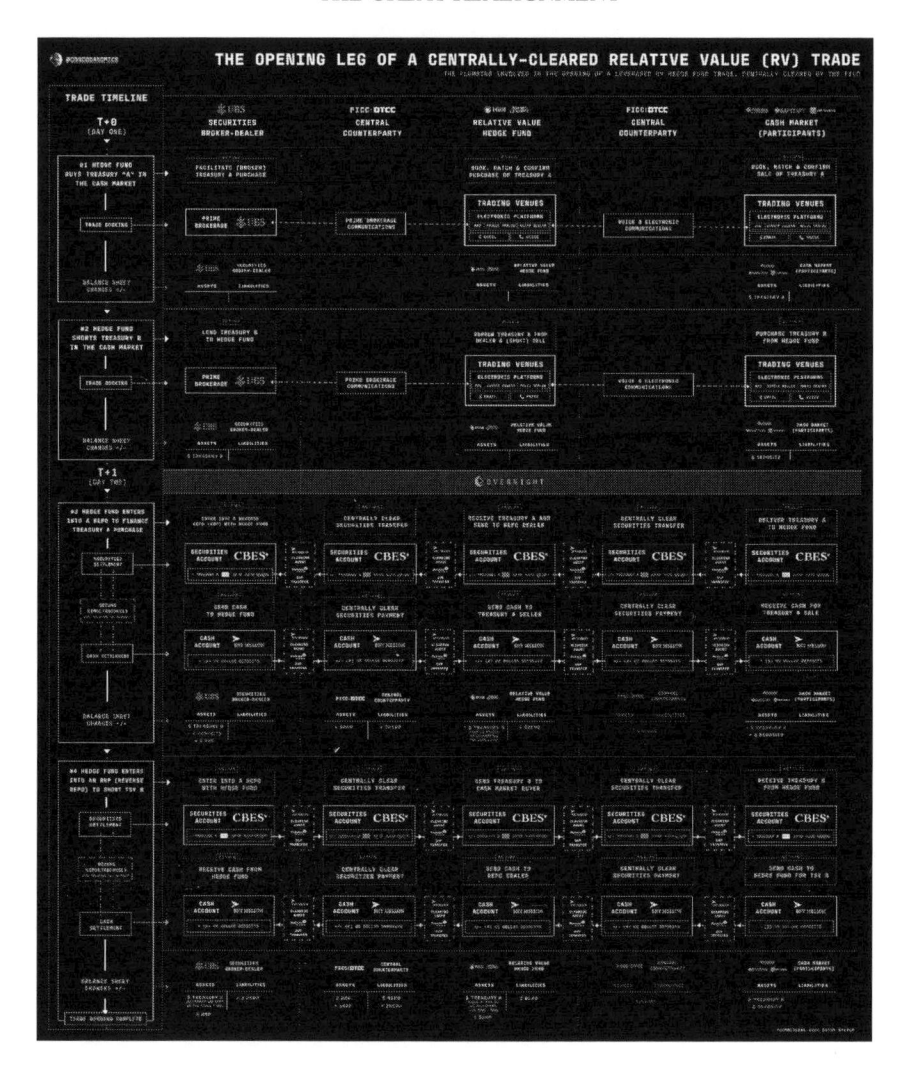

I would wager that the number of people that can make heads or tails of that chart, or indeed, find the exercise even remotely tolerable, would probably number in the thousands worldwide. And this is actually a relatively straight-forward process in the grand scheme of the global financial machine. When I say there is enormous human thought, effort, energy and momentum embodied in the political economy, this is only a tiny fraction of what I am referring to. The act of coordinating global trade, settling accounts, creating money, moving commodities, issuing currency, settling foreign exchange and all the other associated plumbing is extraordinarily complex. For us, the mere end-users, it is all pretty simple. We present a card or fork over some paper and we get things. We show up at an office, we send emails, we flutter about the water-cooler, and people fork paper over to us. Maybe, if we are really "sophisticated," we hop on an online brokerage or sit down with our HR manager to invest some money in a market fund. The depth of understanding about how all of that gets accomplished is largely left to the realm of magic. Or, at least, it may as well be.

So much so that the vast majority in the western world simply take it all for granted. As I pointed out in the section on Power, all of this complexity gives rise to the type of chaos that allows for ascension up the greased pole of power in the modern economy. Those that are best equipped and able to coherently navigate this enormously complex system are the ones that eventually find themselves in the competitive selection process that results in becoming an authority. That said, the idea that nothing more than naked greed can fuel the ambitions of these people is a bit of a stretch. Much like the nobles of ages gone by, many of these people see themselves as destined for this work. And given the broad assumption that "life is a competition for scarce resources" where only the "cream rises to the top," when these competitors do rise to each successive tier they, naturally and predictably, end up falling victim to their own hindsight biases. This is what makes Jeff Bezos' admission about "just getting lucky"[376] so interesting. Not because it is an accurate assessment. It is rather because he has the self-reflection and a constitution that allows that thought to exist at all. This is an exceedingly rare trait among the survivors of selection processes that require very rare successes in very complex competitions.

Bearing in mind, of course, that the progenitors of the financial crisis are merely cogs in the greater game of modern competitive resource capture. Going beyond the realms of finance lie the worlds of fashion, agriculture, entertainment, electronics, automobiles, food production, the list is virtually endless. Meaning, woven throughout the entirety of the Western economy are millions of iterations of similar processes dominated by similarly overconfident decision makers. The readily observable outcome of these millions of iterations is a level of waste and material squander that is truly hard to fathom. Half a trillion dollars[377] worth of food is wasted every year in the United States alone. Planned obsolescence, as an industrial practice, dates back to the 1920s.[378] Whether this line of reasoning was a natural by-product of the industrial revolution, or was incentivized by the advent of fiat currency, or some mixture of the two would be near impossible to reconcile. Regardless, the result is an extraordinary waste of natural and human resources in the never ending pursuit of profit. But without that waste and without those profits, there would be no work for everyone to do. Without that work, without that profit and, indeed, without that waste, the great game of resource capture would slowly grind to a halt. It should come as no surprise the largest economy on the planet is also the <u>most wasteful by a wide margin</u>.[379]

To say this is all somehow driven by greed alone is not a defensible statement. It is certainly not lost on the observant that those who most publicly rail against corporate greed are the same people that feel quite comfortable to live within its warm bosom of waste and destruction. They wear fast fashion, they own cell-phones, they use computers, they walk on paved roads and live in houses. None carry water drawn from a polluted stream back to their mud hut in which their children lie suffering from malnutrition. In their anguish and rage against greed, their own biases allow them the fiction of advocacy while being the direct beneficiaries of these wasteful practices. But to ignore, or worse yet, to reify the resultant waste and destruction is also indefensible. To assume that it is all well and good that we create endless mountains of garbage while allowing the global majority to go hungry and live in squalor is outrageous. To say that this competitive system has endowed us all with a righteous place at the top of the hierarchy which affords us an endless variety of choices while the small folk starve is unconscionable.

Yet the trajectory of progress cannot be denied. Prior to the pandemic, the number of people living in extreme poverty had been <u>reduced by 50%</u>[380] between 2010 and 2019. As the forces of globalization pushed outwards, millions of people around the world saw their standards of living rise, their health metrics improve, and their <u>child mortality rates plunge</u>.[381] Of course, this was not uniform progress. Some areas improved much faster than others. Many areas saw declines, and in the worst cases, states failed altogether. The overarching issue and, indeed, the looming trouble with this progress is the fact that it was all created and birthed by debt. As noted earlier, the entirety of the global economy rests on a base-layer of sovereign debt. The resource to be captured is sovereign debt. The incentives are born of sovereign debt. The progress achieved in the global political economy is erratic and uneven largely for that reason. And it is largely for that same reason this progress will naturally retard and eventually must reverse.

"Never attribute to malice that which can be adequately explained by incompetence or stupidity."

The modern political economy is no more being driven by greed than it is being driven by those that are "the source and motivation to create wealth." It is merely the latest refinement in a game born of deeply flawed starting assumptions. This particular iteration mandates that debt pay for all progress just as it incentivizes the waste required of that progress. It can work no other way. The players operate within this construct, not necessarily out of naked greed, nor with any particular acumen for success. As stated at the outset, unlike power and money, greed is not an abstract concept. It is not a human designed and created system. It is a social judgment with problematic subtexts.

The point of this section is to highlight some of those problems. The purpose is to peel back the layers to reveal broader view of the complexity in the modern political economy. It is an entirely reasonable thing to march about the world with a certain degree of certitude about the nature of our existence. This manner of functioning is required to navigate physical and social spaces. All of us would be rendered to incapacity if we endeavored, moment by moment, to quantify every interaction with every molecule we encounter. Yet when discussing complex social dynamics, the mental short-cuts that enable us to exist as a species can often prevent us from seeking, and sometimes even block us from achieving, progress in the human condition. Rigid adherence to dogmatic ideas is a time-honored recipe for stagnation of ideas.

The great danger to us all is the desire to succumb to simple solutions for the enormously complex issues we all face as humans writ large. One of the many beauties of Bitcoin is that it is at once both simple and complex. Bitcoin, as a neutral arbiter removes much bias from the social interaction equation. In this book my aim is to highlight the human biases and, often unfounded, assumptions that have guided society thus far. Humanity has made great strides in understanding the physical universe. My fear is we have not come quite so far as we have led ourselves to believe when it comes to the social sphere. It is quite normal for people to compartmentalize humanity through time. Through the time-based fictions of epochs, and ages, and generations, we allow ourselves the deceit that those who came before us were somehow separate from us, solely based on their timeline. The reality is quite different. History is an unbroken narrative. People today are governed by nearly identical biological and social processes as those that lived 50,000 years ago.

We have all been playing the same game of competitive resource capture that our forebears started some 12,000 years ago. The clothing has changed. The language has changed. But there is a compelling reason why history may not repeat, but that it certainly rhymes. Five hundred years from now, the people then will likely look back on us and think us to be as backwards and uninformed as we look back on the people from the 1500s. Yet this assumption of backwardness is yet another bias and error. If time travel were possible and we could go back and share what we know and who we are, those folks would catch up much faster than we might give them credit for. Indeed, in 2012, a group of researchers did something functionally equivalent. They went to remote African villages and delivered Motorola tablets to the kids in the village. These children had no exposure to written language—not even a road sign. They certainly had no exposure to computers and the internet. With no instructions, not even a label on the box, the following experience[382] was reported by one of the researchers:

> *I thought the kids would play with the boxes. Within four minutes, one kid not only opened the box, found the on-off switch ... powered it up. Within five days, they were using 47 apps per child, per day. Within two weeks, they were singing ABC songs in the village, and within five months, they had hacked Android.*

I think this does not speak well of our relative advancement. It rather demonstrates, as far as I am concerned, that our relative sense of human progress is not as far along as we might like to believe. Sure, we have electricity and computers and all that fun stuff. But the vast majority of us, including those in leadership positions, are not the architects of this progress. Indeed, most of us are not even the maintainers of it. Much like the kids in the African village, we are, by and large, end users and consumers. The architects have just done such an amazing job in making it all readily accessible and understandable. Even to those who are, ostensibly, "backwards" from us. The point of this book is to help unpack the idea that we may be doing ourselves a collective disservice by assuming our understanding of the global social environment has meaningfully advanced simply because our tech has gotten better. The very notion that our understanding of history, economics, psychology, sociology and biology is distributed amongst our collective consciousness as thoroughly as our gadgets are. "Bitcoin fixes this" is a common trope among the faithful. It also assumes we all collectively know what needs fixing. The argument I will make in the final part of this book is that the fix Bitcoin brings may be very different from what we might be hoping for. Indeed, it may even be better than we can imagine.

PART FOUR: BITCOIN

CHAPTER FOURTEEN

As you have likely gleaned from the title, this book also features a discussion about Bitcoin. Unlike the first three parts of the book, however, the Bitcoin discussion is not about the past. It shall rather, and largely, be forward looking instead. For my part, I first got into Bitcoin in 2011. I tried and could not figure out how to mine Bitcoin. I was able to locate someone on a forum who agreed to sell me some. If I remember correctly it was about $100 worth. Shortly after I received my Bitcoin the price dropped by more than 50%. It was just bad timing. But I was skittish enough to write the whole thing off as a scam and deleted the wallet. People sometimes ask if I can somehow get access to that Bitcoin or if I regret deleting the wallet. I do not. I never have. I know myself well enough to know that I would have sold that Bitcoin a long time ago. Certainly long before it ever hit $2000, let alone $70,000. The trouble is, I tend to be very practical. And, as a life-long Libertarian, "practical" meant gold and silver, not some magic internet money. I continued to be very "practical" about Bitcoin for nearly a decade. In 2016, I wrote a multi-part series on anti-money laundering (AML) laws for the *Banking Law Journal*.[383] Within that lengthy 60-page article series I devoted about three pages to Bitcoin. Even then I did not return to the Bitcoin experiment beyond being a casual observer for another four years.

Nevertheless, before we press on with the discussion on Bitcoin, I must say I am operating under the, perhaps mistaken, assumption that most people reading this will already be familiar with Bitcoin. On the off-chance that is incorrect, I will provide a quick primer on the system. I am lifting this straight out of my paper on money laundering.[384] I do this for the sole reason that it was written for lay audiences and I think it does a fair job of covering the core functionality of Bitcoin. If you want to learn more, there are a number of sources devoted to the topic. I would recommend Matthew R. Kratter's *A Beginner's Guide to Bitcoin*[385] for an excellent introduction. That said, while there are numerous variations on digitally based currencies, the most widely used is Bitcoin. Bitcoin is a decentralized virtual currency that is based on a peer-to-peer (P2P) network distribution model. The central functionality of the system relies on a public ledger called a blockchain. This ledger is continually updated with all of the transactions that occur between Bitcoin users. This ledger is also decentralized and is "managed" by users that run Bitcoin nodes. In order to ensure that transactions are authentic, Bitcoin uses a novel concept based on asymmetric cryptography. Under this system, users are assigned two linked and numerically unique keys—a public one known to everyone and a private one known only by the user.

In essence, the user's public key provides a record of transactions made by that user, whether sending or receiving. The private key is necessary to actually send funds or to retrieve them. Thus, once funds are sent, only the recipient can access them via their private key, preventing retrieval by the sender. With a publicly accessible and viewable log of the transaction, the receiver cannot claim the funds were not received, because the record would reflect affirmative access to the system via the sender's private key. As previously noted, the system is decentralized, so there is no entity in the middle to verify ledger entries. Rather, the system verifies transactions via users called "miners" that devote computing resources to calculating the hash values of the transactions, with correct proofs adding to new blocks of data in the ledger, thus the blockchain. Transaction verification is not the only function of the miner, as their computing resources are also devoted to the creation of new bitcoins (the currency unit of the system). The rewards for the miner are comprised of fees paid by users to include transactions in a block and receiving new bitcoins for providing a correct proof for a block sequence. Higher fees are sought by miners, which is how users can encourage their transaction to be prioritized into the next block being calculated. Once a transaction has been initiated, for the sender and receiver to have full confidence in the transaction, they must wait until the proof is computed. This typically happens within ten minutes, but can take several hours depending on the fee pledged and the volume of network traffic.

While Bitcoin is often touted as an anonymous payment system, just by the fact that the transactions post to a public ledger means that a record exists. There are varying means by which a user can obscure the digital trail, such as splitting off bitcoins into smaller amounts and depositing them at multiple addresses. However, doing a heuristic analysis of publicly available keys and the associated transactions, researchers from the University of California at San Diego[386] (USCD) discovered that they could trace public keys utilized in a transaction to exchanges that are subject to government authority, i.e., physical institutions where the authorities could deliver a subpoena or serve a warrant. While the researchers point out that individual transactions could be hard to identify, for someone to put a large amount of bitcoins to practical use will eventually require the services of an exchange. The UCSD researchers do acknowledge that the adoption of more rigorous protocols could serve to increase the anonymity of the transactions and make bulk movements harder to detect, but to do so currently requires a significant effort. Even still, unless and until Bitcoin or other alternative currencies that lack sovereign support are widely accepted, any potential user will, at some point, still be confronted with the need to convert the exchange medium (i.e., bitcoins) into fiat currency or other tangible items of value in the "real" world.

Keep in mind, the summary you just read was written nearly ten years ago. Regarding the latter properties of transactional use, there is a burgeoning use for Bitcoin as a routine transactional medium. However, there remain, to this day, barriers to the widespread adoption of Bitcoin as a medium of exchange. Fiat denominated price volatility is problematic for settlement unless both parties compensate or create terms that account for fiat denominated price fluctuations. Moreover, transaction speeds[387] on the Bitcoin network are not sufficient to accommodate a high transaction volume that can reliably sustain even modest trade. There are so-called "Layer-2" network solutions[388] that address the transaction speed issue. The "Second Layer" in the name refers to an alternate network that processes and / or batches transactions with only periodic settlement to the "base" layer—the actual Bitcoin blockchain. All have various trade-offs in terms of security, tax implications³⁸⁹ and centralization, or the need for trust in third-parties. In regards to the privacy of transactions noted in the last paragraph, the tools to track Bitcoin transactions³⁹⁰ have gotten significantly more sophisticated since that was written. Modernly, and practically speaking, to reliably obscure beneficial ownership of Bitcoin and then be able to transact with that obscured Bitcoin anonymously requires a high degree of technical savvy possessed by very few.

Put simply, as a medium of exchange for routine transactions, Bitcoin is currently a poor choice. This is especially true if your anonymity is of critical importance, e.g., you are a freedom fighter under an oppressive regime. It must be said, however, that fiat currencies also suffer from these same issues, albeit they only generally become noticeable to end users when conducting international exchange. They become especially acute when conducting large scale transactions between sovereign jurisdictions. One of the easier industries to illustrate this issue with is the international film industry. While most people likely do not stop to ponder the inner workings of the film industry, if you are a film production and distribution company, foreign exchange market³⁹¹ (FX) volatility is highly problematic. For background, most film production companies license the films³⁹² they make to play in theaters overseas. Those license deals are generally predicated on taking a percentage of box office receipts in exchange for the rights to play the film. The problem with this is the fact that foreign markets do not sell movie tickets in U.S. dollars. Their box office proceeds are collected in their respective local currencies. The trouble arises when trying to determine the timing for the payment of the license fee. If a film does very well at the Japanese box office, for instance, and the license fee remittance is due 90 days after the release window, the Japanese Yen will not likely exchange at the same rate as when the deal was closed. The Yen to U.S. dollar exchange can vary widely within a 90 day window. The contracts utilized³⁹³ to settle these deals are extraordinarily complex. Even with that complexity, conflicts and lawsuits abound as each respective side tries to maximize their gains and minimize losses by playing fast and loose with the settlement agreements and payment timing.

Likewise, fiat currency also suffers from anonymity problems when conducting transactions overseas as well. Anyone that tries to wire a significant sum of money overseas is probably aware of the identification requirements and other issues that can arise from transacting internationally. In fact, one of the preferred methods for international drug cartels to launder money is via the rather crude form of bulk cash smuggling.[394] For them to reliably retain large scale transactional anonymity, they simply pack a couple million dollars in a backpack and pay a network of "mules" a few hundred dollars each to walk it across the border. Once the cash is in a friendly jurisdiction, it is a simple matter[395] for them to integrate that money into the international finance system through shell corporations and with the assistance of compromised bank employees. In terms of transaction speed, Bitcoin is certainly more efficient than an international wire, as those transaction speeds are administratively throttled by the sovereign monetary control apparatus. That throttling makes Bitcoin's roughly half-dozen transactions per second[396] seem lighting fast by comparison. Put another way, Bitcoin as a transactional medium is, for all intents and purposes, no different than any other sovereign issued currency, save for the fact that Bitcoin has no sovereign. With that lack of sovereign issuance Bitcoin must, of a necessity, "float" on the market just like all the other currencies. The problems that arise using Bitcoin for exchange are no different than any other monetary unit. They are just more readily apparent to the end users because, unlike sovereign issued currencies, those problems are acute both within and beyond sovereign borders. To extrapolate this phenomena out a bit, each Bitcoin holder is effectively a monetary sovereign unto themselves. Thus it is unsurprising that they would be subject to the same market fluctuations other sovereigns must navigate when conducting inter-sovereign currency exchange.

I mention all of this up top for the very simple reason that the Bitcoin Whitepaper is titled in full, "Bitcoin: A Peer-to-Peer Electronic Cash System."[397] The author of the paper makes numerous references to proposed utilizations of Bitcoin that are analogous to cash transactions. For example, on page one it is clearly stated, "What is needed is an electronic payment system based on cryptographic proof instead of trust, allowing any two willing parties to transact directly with each other without the need for a trusted third party." Pausing for a moment here, let us recall the definitions of money provided in the section on Money, which are:

1. A medium that facilitates the exchange of value in low, or zero-trust situations; and/or

2. A medium that allows for the bulk extraction of fractional value from the personal or collective industry of sovereign subjects through taxation; and/or
3. Is an asset that appreciates in nominal price at, or near, the same rate as the sovereign increases the quantity of the first two.

While it is an often debated point, it seems rather clear to me the original intent of the creator of Bitcoin was largely to satisfy prong one of the definition above. Regarding prong two, clearly Satoshi Nakamoto was not interested in creating an efficient digital taxation device for national sovereigns. It is also clear from Nakamoto's email and forum correspondence[398] that Bitcoin becoming an asset that appreciates in nominal price had some consideration as well. To wit, in 2010, Nakamoto pondered a thought experiment in a Bitcointalk forum post[399] regarding a scarce metal similar to gold, but with a "magical" transmission property. Of this magical metal Nakamoto writes:

> *If it somehow acquired any value at all for whatever reason, then anyone wanting to transfer wealth over a long distance could buy some, transmit it, and have the recipient sell it.*
>
> *Maybe it could get an initial value circularly as you've suggested, by people foreseeing its potential usefulness for exchange. (I would definitely want some) Maybe collectors, any random reason could spark it.*
>
> *I think the traditional qualifications for money were written with the assumption that there are so many competing objects in the world that are scarce, an object with the automatic bootstrap of intrinsic value will surely win out over those without intrinsic value. But if there were nothing in the world with intrinsic value that could be used as money, only scarce but no intrinsic value, I think people would still take up something.*

I would be remiss to not again point out that, with the possible exception of life sustaining necessities, e.g., water, air and nutrition, nothing has an "intrinsic" value to human beings. Value is irrevocably and irretrievably a context dependent, and entirely subjective metric. This is fundamentally true despite any contrary insistence by the gold-bugs of the world.[400] I would also point out that Nakamoto is correct. Which is to say, Nakamoto is correct if we hold as true that nothing can have "intrinsic" value. In that void, people have obviously taken "something" up as money—those "somethings" being things like gold, silver, and paper. Regardless, within this book, I place little emphasis on the utility of Bitcoin as a routine transactional medium. The ability to buy a cup of coffee, or even a bag of cocaine, with Bitcoin appears to me a triviality in the grand scheme of things. I recognize many very thoughtful and intelligent people may disagree. Indeed, Nakamoto themself may well disagree. With that in mind, let us trace back for a moment to the earlier discussions on Money and Power. If we assume, and I do, that competitive resource capture economies under sovereign control will continue to dominate global trade and interactions, then it seems unlikely those sovereigns would abandon a central component of their authority, i.e., their money. Which is partly why I also place little emphasis on the notion of a Bitcoin-backed sovereign currency. Beyond the loss of sovereign control over monetary policy, as longtime Bitcoin developer Eric Voskuil correctly points out,[401] for the end users, there would be no meaningful way to audit paper currency issuance versus Bitcoin held. If there were a way to verify and limit issuance, then the currency units would be indistinguishable from the underlying Bitcoin asset itself. The issue there being, if the sovereign decides to expand the number of notes circulating beyond reserves of Bitcoin available for redemption, there is no mechanism to prevent that from happening without essentially duplicating Bitcoin's core functionality. Without such a limitation, a Bitcoin reserve currency would inevitably end up in the same conundrum that collapsed Bretton-Woods.

All the same, the focus here is rather on the third prong of the money definition provided. Which is to say, the focus instead shall be on Bitcoin's properties as an asset that appreciates in nominal price at, or—in the unique case of Bitcoin—superior to, the rate the various sovereigns debase their paper currencies. Among those sovereign currencies, special attention will be paid to the U.S. dollar. Before we dive in, I think it would be helpful to briefly explore and discuss other digital currencies. While awareness of Bitcoin has certainly expanded dramatically in the last few years, for the lay person in particular, I think it is very difficult for them to distinguish Bitcoin's unique characteristics from other distributed ledgers.

Moreover, sovereign nations are now experimenting with digital currencies of their own. These so-called Central Bank Digital Currencies (CBDCs) provoke a number of widely varying reactions among advocates and opponents alike. The goal here is to—hopefully—dispel some of the myths and orient the conversation in a more constructive direction. In turn, this will also—hopefully—prove helpful in the later discussion about Bitcoin as a non-sovereign treasury asset.

CHAPTER FIFTEEN: ALTS AND CBDCS

So-called "alt" coins are, generally speaking, distributed ledger based schemes that ostensibly provide an alternative to Bitcoin. The colloquial and catch-all term for these "alt" coins is crypto, or cryptocurrency. Bitcoin is often characterized as crypto as well, though Bitcoin proponents, myself included, are quick to distinguish Bitcoin from "crypto." As the first cryptographically secured distributed ledger, Bitcoin has a number of unique characteristics that distinguish it from the 15,000+ "alt" coins[402] in existence. Chief among them is the means by which the developer created and distributed Bitcoin. Satoshi Nakamoto is a pseudonym for an unknown person, or group of persons, that created Bitcoin. The Bitcoin concept was published via the Bitcoin Whitepaper,[403] while the Bitcoin code was released for free as an open-source project.[404] Thus, anyone with the capability to develop, improve, or even copy and modify the system is free to do so. As this network of participants and developers grew, Satoshi Nakamoto simply disappeared.[405] This lack of an identifiable entity upon which an adverse interest could focus attention is significant for a number of reasons, not least of which is the inability for a sovereign to have someone or something to capture or coerce.

Bitcoin also enjoys an enormous first-mover advantage.[406] This is partly evidenced by Bitcoin's current market capitalization, especially when compared to the rest of the crypto market. The current total market capitalization for Bitcoin and cryptocurrencies combined is approximately $2.2 trillion. Bitcoin alone commands more than half of that sum. Put differently, Bitcoin is the market. With what remains, Ethereum runs a distant second, while the other 15,000+ cryptocurrencies fractionally split the rest. Moreover Bitcoin's network effect[407] is currently of such a magnitude that it appears highly unlikely a competitor upstart will displace it. To be certain, there have been several attempts to capture Bitcoin's network effect via hard-forks of the Bitcoin code.[408] The vast majority of cryptocurrencies created in the wake of those attempts have rather pursued a "similar, but better" angle through novel code implementations[409] with expanded functionality. This is yet another area where Bitcoin is distinguishable from cryptocurrencies. As noted above, new bitcoins are created from "mining" operations, where correct transactional proofs are rewarded in part with new bitcoins. This is analogous to physical mining or drilling operations, where correct extraction methods are rewarded with "new" sources of commodity supply.

When Bitcoin was launched, there were no bitcoins in existence until users devoted compute resources to create them. No market existed for bitcoins until an early user facilitated a trade of 10,000 bitcoins for two pizzas.[410] By way of contrast, one of the more successful cryptocurrencies is Ethereum.[411] Like Bitcoin, Ethereum started life as a mined crypto. However, unlike Bitcoin, the developers of Ethereum both remained in the public eye and created a fund-raising apparatus[412] to boot-strap the endeavor. In addition to selling "Rep" tokens pre-launch, around 72 million Ether[413] (the currency unit of the system) were subsequently "pre-mined," meaning they were mined by the developers in advance. These pre-mined tokens were sold to the general public in an initial coin offering (ICO). While the exchange was for Bitcoin, the initial fiat equivalent price of each Ether was approximately $0.30, which climbed to nearly $3.00 in early trading. Roughly 10 million of those original Ether were reserved for the developers themselves. Seven months after the launch, Ether was trading for more than $12—an overnight windfall for the developers, who essentially conjured the tokens out of thin air. Modernly, after several hard-forks, complete re-writes to the code, and numerous other changes implemented by the Ethereum Foundation, the privately operated, company branded and trademarked software token[414] called Ether trades for around $2600.

That early ICO windfall was not lost on the broader community of burgeoning crypto developers. Indeed, a veritable cottage industry[415] sprang up around the ICO model. The quality of these projects were, and are, generally quite poor. To this day, they all share many of the same characteristics as the companies that drove the so-called "Dot-Com Bubble"[416] at the turn of the century. Wild promises of tech-driven transformative potentials abound, cloaked in technical jargon, while the underlying fundamentals of revenue, profit, and expenses are largely ignored or obscured. For instance, the popular distributed ledger platform Solana touts high transaction throughput[417] and fast time to transaction finality as transformative potentials. The retail narrative posits that Solana's technology will revolutionize everything from online gaming to finance. These promises have helped drive Solana to a market capitalization of $68 billion, most of which is represented by the monetary premium created by speculators in the Solana token. All this despite a meager $282 million in annual revenue,[418] with more than ten times that in annual expenses. This leaves Solana with negative earnings totaling hundreds of millions a year. For comparison, the Nintendo Corporation[419] has a $64 billion market capitalization on revenues of $11 billion[420] and earnings of over $4 billion.

Even so, research done by Satis for Bloomberg revealed that more than three-fourths of all ICOs were outright scams.[421] The vast majority that were not scams were either abandoned or scarcely functioning a year post-launch. Much the same is true today, as the public facing narrative is to tout high-tech innovation, while behind the scenes, the functional business model is to fleece would-be retail investors.[422] The U.S. Securities and Exchange Commission (SEC) has filed and won numerous lawsuits[423] against a number of crypto projects. However, the sheer volume of fly-by-night crypto ventures, coupled with the limited resources at the SEC,[424] tilt the risk-to-reward ratio strongly in favor of the crypto developers. As currently staffed, the SEC is simply not equipped to meaningfully enforce securities laws against crypto developers, many of whom are domiciled outside of the United States. Make no mistake, however, that the overwhelming majority of crypto projects operating today were launched in a manner that runs afoul of very clearly established securities regulations[425] in the United States. While crypto proponents claim a lack of regulatory clarity, these rules pre-date WWII and there is no carve-out for "shiny and new." Meaning, despite even well meaning arguments to the contrary, with very few exceptions, the funding models for nearly every noteworthy crypto project launched in the last decade involved the illegal sale of unregistered securities.[426]

Bitcoin, in yet another contrast, is not a security. Bitcoin is a commodity[427] and, for tax purposes, is treated as property[428] in the United States. Notably, it is not construed as a currency in either case, despite the ability to conduct transactions solely with, and denominated in, Bitcoin. To quickly sum the major contrasts between Bitcoin and other digital ledger schemes:

1. Bitcoin has no issuer
2. Bitcoin is decentralized
3. Bitcoin has an exponential network effect advantage
4. The Bitcoin market is a laissez-faire[429] creation
5. Bitcoin (currently) has legal clarity in the United States.

To fully expand upon the intricacies of the crypto world would be another book unto itself. The short version is crypto is largely a vehicle by which developers and their venture capital partners enrich themselves at the expense of uninformed retail investors. There is an enormous influence apparatus that supports these operations. All of these operations and support functions are currently illegal in the United States.[430] It is against the law to raise start-up funds from retail investors without registration with the SEC. It is illegal to promote those schemes without registration with the SEC. It is illegal to engage in market manipulation.[431] It is unethical to engage in coordinated pump and dump schemes.[432] Indeed, a pump and dump becomes illegal if the developers and early investors are aware of this occurring. It is also illegal to create a project and then steal user funds,[433] which as noted earlier, comprises more than three-fourths of crypto projects created. While crypto developers and their venture capital partners complain endlessly about regulatory uncertainty, they continue to engage in illegal activity for the sole reason that it is highly profitable and hard to stop.

This all despite the fact that the SEC greatly softened rules[434] regarding crowd funding start-up capital in 2016. The CROWDFUND Act, quite literally, paved the way for every single one of these entities to legally, and for relatively low-cost, register their offerings, provide appropriate disclosures, and still raise money for their proposed endeavors. Meaning, fully one year before the "ICO craze" of 2017,[435] all of these bad actors had a very easy and straightforward path[436] to legally raise funds. Not one of them took advantage of this legal path. Instead they chose to "move fast and break things"[437] with no regard for the human toll they extracted. Their retail victims have lost, and continue to lose, billions of dollars investing in their faulty schemes,[438] with near zero recourse. And, to this day, these bad actors continue apace, desperately trying to muddy the waters. They go to great lengths to dilute the distinction between their naked scams and Bitcoin. Let it be said in no uncertain terms, Bitcoin is not crypto. Bitcoin is a unique digital asset that has no parallel in the history of the world. A bold statement to be sure. And it is one I intend to develop and support in the following chapters.

Central Bank Digital Currencies

Few things get freedom loving people more riled up than the thought of a central bank digital currency (CBDC). Once cash goes away, they say, you can kiss your privacy and freedom goodbye. Fear of CBDCs being used to enforce Chinese-style "social credit" scores, "carbon" scores, or vaccine mandates probably top the list. The loss of financial privacy also ranks highly among the fearful. This fear takes many forms on social media:

Arguably, many of these fears are overblown. This is, however, not to say they are unfounded.[439] CBDCs, as most envision them, simply remove cash. Once cash is removed, as the logic goes, then Big Brother[440] can enforce all manner of draconian practices against the population. Given the history of governments over the last few hundred years, it is a very reasonable fear. However, the United States was born out of frustration with the abuses of the so-called "ruling" class. At its core, the United States laid down a set of rules that recognize inalienable rights. Life, liberty and the pursuit of happiness comprise the nut-shell version. The more thorough one says government authority is delegated and restrained, with ultimate authority residing within the population writ large. Of late, the value of those paper promises have proven somewhat less than resilient.[441] Nonetheless, there is an in-principle guarantee of certain freedoms and rights within the United States. Meaning, unlike our Eastern neighbors in China, there are many barriers for would be Supreme Leaders[442] to overcome before a Chinese style "social credit" system could be implemented domestically in the U.S.

Before we continue, let us briefly recap earlier chapters. Recall that post-agricultural Western European governance essentially evolved from roving groups creating protection rackets through coercive violence. Big, strong men would band together and basically pillage their way through the world. Occasionally these groups would fight each other over who got to pillage which place. Much like street gangs today, they would lay claim to an area and defend it with force. Within the respective group's area, they could extort and pillage as much as they desired. Drift into another group's territory, however, and they would be met with violence. If they succeeded in the tournament style competition for those resources, they would claim that territory to increase their ability to extort and pillage. If they did not prevail, then the other group expanded instead. On and on it went through pre-history until a rough equilibrium of force was reached. Smaller groups were incorporated into larger groups. Hierarchies formed. Territories expanded. After a few thousand years of refinement, what started out as simple extortion has morphed into our modern concept of the nation state. This all led to the primary component that makes a nation "sovereign"—the authority to coercively use force within their respective sphere of influence. Modernly, the use of coercive force absent a sovereign designation is near universally a crime. Meaning, I cannot legally force you to pay me a tax, nor can I force you to labor for me—nor can I force you to do anything else for that matter. In all cases, and in all countries, that power is solely reserved to the sovereign, whether they be a king, a queen, a dictator, or an elected body.

Likewise, and as noted previously, people define "money" in all manner of ways. Unit of account. Store of value. Medium of exchange. All definitions work to some degree. The trouble is, none of them adequately capture what money truly represents. Yes, money is used for those things. But, as argued in the section on Money, that is not the core reason it is there. The reason provided for money existing in the form we use today is essentially an abstraction of sovereign force through the vehicle of taxation. Put another way, if we trace back to those roving groups of strong men in pre-historical Europe, their modus operandi would simply be to go into a village and steal what they want. However, as these groups (gangs?) grew in number, they ran into a logistics problem. Of note, every participant in these groups would rightfully expect to be paid for their contribution to the looting. When these groups were small in number, then dividing up the spoils of extortion is relatively straightforward. But when there are thousands of group members, paying them in chickens, pigs or corn might ensure they are compensated right now, but what they do not use immediately will either spoil or go to waste. This same conundrum is, likewise, true of the sovereign at the top of the chain. Their conundrum is only magnified. The way around this dilemma for these enterprising group leaders—kings and queens and the like—was to force their subjects to pay for their extortion with monetary units created and issued by the sovereign. Thus, instead of proffering a chicken, or a pig, their subjects would have to figure out how to get sovereign issued money. Meanwhile, since the sovereign wields a monopoly on the use of force, they can also force their subjects to accept money from all their other subjects—thus, the birth of the "legal tender" law. Recall, the net effect of this scheme is that:

1. Sovereign subjects must use sovereign money to pay the sovereign; and
2. They must accept sovereign money as payment for everything else.

The way they procure sovereign money is by trading their time, labor, and/or resources in exchange for that sovereign created money. One of the niggling problems with this scheme is counterfeiting. If sovereign issued money is easily counterfeited, then the whole racket falls apart. Which is one reason why gold and silver were so commonly used early on. Gold and silver can be reliably assayed.[443] While it is inconvenient to assay a metal, it is convenient enough to tell if someone is tinkering with your money at scale. Moreover, gold and silver already had a long history of perceived value, was relatively easy to work with and relatively scarce when refined. Put together, little coins made out of gold and silver became a great way for sovereigns to loot their population without having to store the tangible goods the sovereign extorted to begin with. The convenient knock-on effect being, the sovereign gains the ability to procure necessary goods and services from their population when they need it.

Once that system was in place, exchanging goods and services became part and parcel of sovereign economies. Likewise, saving, lending and borrowing money becomes an industry in and of itself. But, at the end of the day, the purpose money is arguably designed for is the extraction of value from a population. The accounting, exchanging, and storing aspect of money all came after that purpose. Be that as it may, using gold and silver as money is still problematic beyond issues of counterfeiting. It is limited by how much has been extracted and refined from the ground. Meaning, if the sovereign—or anyone else, for that matter—wants to increase how much gold or silver they have, they either need to find more to mine and refine, or they have to acquire some from someone that already has it. Bearing in mind through it all that sovereigns are, broadly speaking, rather poor at managing money. The trouble is, people are involved. And, whenever people are involved, they tend to orient towards actions that maximize rewards, while minimizing effort. Spending more money than one has is an all too common example. Thus, even when silver and gold were money, this did not stop the sovereign from spending more than the sovereign coffers held. What they would try to do is overspend and then debase their currency by mixing in pot metals, or by reducing the weight of the coins. This way, the sovereign could make it appear as though they had more money than they really did. It is counterfeiting if anyone else does it. But, since the sovereign is the only one that can use force, there is a major barrier in place for those who would try to stop this practice.

That said, sovereign counterfeiting schemes would work for a period of time. Which is to say, they would work until everyone figured out the scheme. Then their subjects would tend to hoard the older, more valuable coins,[444] and use the new debased ones to pay taxes and buy goods. The introduction of paper promissory notes thus became a way for the sovereign to more easily debase their money. The way the scheme works, the sovereign simply says, "Dear subjects, now instead of gold and silver, you must trade and pay taxes with these little pieces of paper instead. Each piece of paper is worth some amount of gold or silver. I promise." Then, so long as everyone does not try to cash out at once, the sovereign suddenly has the ability to "make" money without having to go through all the trouble of finding, mining, refining, or stealing more gold and silver. Perhaps unsurprisingly, most European nation-states adhered to the same scheme. And, provided they were not engaged in yet another tournament style competition (war), they were all trading with each other across their respective borders. Largely owing to the fact that they are all engaged in a competition for resources, they inevitably tried to cheat each other by manipulating markets and playing fast and loose with settlements. The cheating and manipulation by sovereigns, both internally and externally, eventually made the various systems break down. Sometimes those breakdowns were limited to one or two countries. Sometimes the contagion spread much further, often with terrible consequences, like global conflict.[445] The last vestige of gold backed paper currency was during the Bretton-Woods era, with the United States serving as the backbone of the system. But, much like their extortionist predecessors, the U.S. sovereign spent too much money. Instead of making it right, in 1970 they reneged on the gold redemption promise they made to the world and went with unbacked paper money instead. While the illusion of prosperity brought many material comforts to the denizens of the United States, once the restraint of an in-principle need to redeem paper for an asset was removed, the debt naturally spiraled. To wit, a visual of just that result:

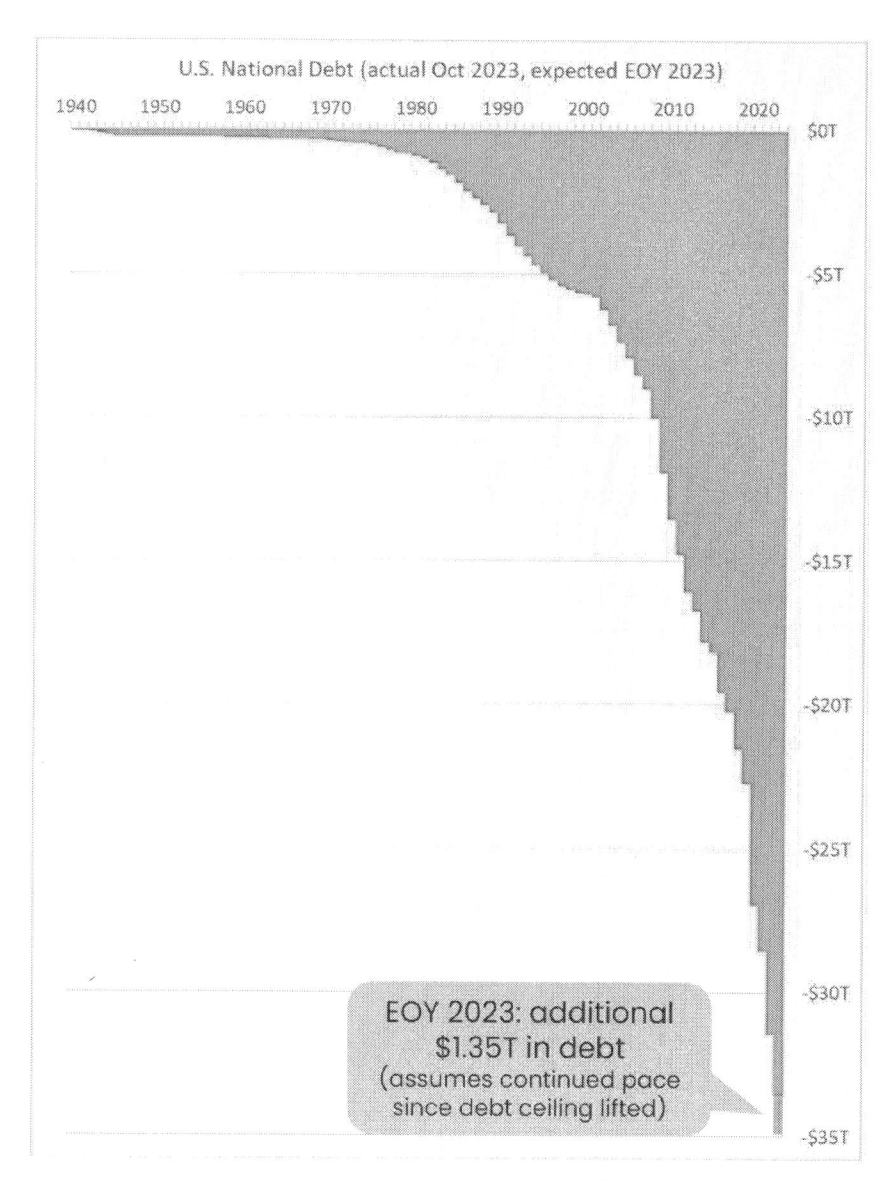

U.S. National Debt (actual Oct 2023, expected EOY 2023)

EOY 2023: additional $1.35T in debt (assumes continued pace since debt ceiling lifted)

With that brief recap in mind, at the core of this problem is the niggling reality that managing a money supply at the sovereign level is hard to do. There are simply too many incentives to cheat, manipulate, over-spend and steal money. And, as alluded to above, the people involved often are forced to make decisions among a selection of terrible options. Moreover, the incentives and rewards that come from maintaining fiscal prudence and being a good steward of money are indirect and greatly delayed. The benefits of sound monetary policy accrue over generations. Those benefits demand, at a minimum, delaying gratification. The base reality is that politics always demands immediate action. As such, "kicking the can down the road" is far more expedient than trying to explain why the can should not be kicked. Indeed, those that try will inevitably fall prey to those seeking power and otherwise lack those compunctions. It is largely for these reasons why a CBDC will likely end up being a rather terrible idea. With that in mind, it must be said that, for the central bankers and Dearest Supreme Leaders of the world, a CBDC does not necessarily represent a tool for oppression. I would go so far as to argue that such a notion does not even rise to the level of consideration in most circles of power. I would posit instead that, by and large, for the Dearest Supreme Leaders, a CBDC is merely an efficiency, much as email is more efficient than sending a letter. That they may end up with a draconian tool for oppression is, rather, a second order effect and potential outcome.

Beyond base fears of CBDCs being utilized as dystopian control tools, opponents also rail against CBDCs because of their potential to infringe on financial privacy. The sad fact of the matter is, any notion of financial privacy was already lost decades ago. Right up to 1970, the only people in the United States that knew where you got your money were the people you obtained the money from. Your bank did not ask, nor care. You could walk into a bank with a million dollars in cash, make a deposit and walk out with a smile, no questions asked. The Bank Secrecy Act (BSA)[446] changed that equation in 1970. The genesis of that act was the widespread use of numbered Swiss bank accounts by stock investors looking to avoid paying taxes[447] on their stock market gains. By modern standards, the BSA was pretty benign. It mandated basic record keeping requirements at banks and financial institutions. The BSA also gave birth to the anti-money laundering morass that we all get to muck around in to this day. Unsurprisingly, when banks were required to ask for ID, people became very angry. The BSA was challenged all the way to the Supreme Court. To that point, the lower courts agreed BSA was an infringement on fundamental rights. The Supreme Court did not agree though, whereby the BSA became the law of the land.

However, the BSA was weak. It was very easy to get around the mandated record keeping requirements. And broadly speaking, cash was still king for transactions in the United States. Meaning, the vast majority of transactions undertaken by U.S. sovereign subjects were still largely private. The powers that be kept trying to wrench down on financial privacy. All through the 1980s and 1990s,[448] they used the "War on Drugs" as an excuse to peer ever deeper into their subjects' personal financial business. In the late 1990s, they went so far as to introduce legislation that would incorporate the FATF-40 in the United States—otherwise known as the Forty Recommendations[449] of the global Financial Action Task Force, which is led, of course, by the United States.

The FATF-40 mandated things like "Know-Your-Customer" (KYC) rules, Suspicious Activity Reports (SARs) and all manner of other invasive practices that substantially degraded financial privacy both in the United States and globally. However, once people became aware of this scheme, there was a major political backlash.[450] That backlash was so intense it caused the legislation to be pulled from consideration. Just a few short years later, the September 11, 2001 attacks gave cover to enact the exact same rules—nearly verbatim—in the elaborately titled "Uniting and Strengthening America by Providing Appropriate Tools Required to Intercept and Obstruct Terrorism Act of 2001."[451] Legislation otherwise known by the more common acronym, the USA PATRIOT Act. An Act, by the way, that could be easily repealed at any time. Meaning, if we truly cared about our financial freedom, it would be a simple matter to restore it. Just repeal the BSA, the USA PATRIOT Act and all the other AML rules that have been passed in-between and since. Which brings us back to the matter at hand—CBDCs and financial privacy. The point of this lengthy exercise is to point out the tools for invasive financial control of the population are already in place.

The FATF-40 effectively deputized the entire banking[452] sector to become state snoops. It allows banks to arbitrarily refuse transactions. It allows banks to seize your accounts under any "suspicion" of criminal activity. Banks already have the ability to refuse to repatriate your own money to you and to refuse withdrawal demands under certain conditions.[453] The U.S. sovereign,[454] the Canadian sovereign,[455] and indeed, numerous other sovereigns[456] throughout modern history have prevented people from accessing their funds.[457] In many cases, the holders of those accounts were debased, receiving only a fraction of the value they held prior to the bank blocking access to their accounts. Moreover, only 16% of transactions are settled in cash in the United States, with the other 84% being conducted digitally. Meaning, 84% of all financial transactions[458] are already conducted via a centrally controlled and monitored digital ledger. A ledger that is exclusively maintained by the sovereign control apparatus. Effectively, all a CBDC would do in the United States would be to shift the remaining 18% of transactions to that same ledger. Keeping in mind, of course, the average cash on hand in America is $369.[459] Put another way, nearly all meaningful transactions that someone might wish to conduct privately are already being monitored. In fact, monitored digital transactions are so ubiquitous today, even prostitutes[460] and drug dealers[461] accept credit cards with scarcely a second thought.

Which is all to say, given the generally low financial literacy,[462] low political participation[463] and general ignorance[464] and apathy of the voting public writ large, when CBDCs finally come along, most folks will likely not care. The ones that do will hoot and holler. They may complain. They may even sue to block the CBDC in court. If the BSA, and other financial privacy destroying regulations like the USA PATRIOT Act are any indicator, those lawsuits will fail. Because of this, I think it much more likely that most people will just forget about it, roll over and take whatever comes. Indeed, the USA PATRIOT Act passed the House 357–66 and 98–1 in the Senate, with hardly a whiff of complaint from the public. Keep in mind, one-third of the USA PATRIOT Act is devoted to implementing the FATF-40. The same FATF-40 the voting public overwhelmingly rejected just three years previous. It did not matter that "money laundering" and "terrorist financing" played no role in the 9/11 attack.[465] In fact, every single piece of AML legislation introduced over the last 50 years has passed with near-zero resistance. Each one that passes is a freedom degrading state overreach that removes more and more financial privacy. People today are so accustomed to being interrogated at the bank they think it is normal.[466] This should already be an unthinkable infringement on the right to free association and freedom to transact. Yet, it is not. It is largely for these reasons that I think CBDCs will be no different. People may whine. They may post nasty tweets on Twitter or angry replies on Facebook. But, in the end, I imagine most people will just take their lumps and move on.

I would, however, be remiss to not point out that the fear around CBDCs presumes that the same people that currently run the global economy will be running the show with a CBDC. Of course, for the reasons noted above, chances are very good that will be true. But, it must be said that things do not have to be this way. A central bank digital currency contains enormous power. If implemented poorly then yes, a CBDC can only be a further degradation. But, the same qualities that make a CBDC a dystopian nightmare for "the people" could also completely constrain and reign in the abuse and corruption of Central Banking if it is programmed in a contrary direction. There is nothing that says a CBDC must be closed-source and programmed to simply replace cash. Practically speaking, every operation a Central Bank undertakes[467] can be hard-coded, immutable and automated with a CBDC. Things like privacy guardrails, limitations on money seizure, transaction tracking, and just about anything else related to private and commercial transactions could be coded in too. If done well, a CBDC could remove human intervention and decision making from monetary policy entirely. Instead, that policy could be written into code, publicly available to audit, and made with no allowance for deviation without an act supported by two-thirds of Congress. The code could very easily dictate the rate of new currency issuance, open market operations, enforce collateral and reserve requirements, and prevent the holding or purchasing of toxic or insolvent assets. No more bailouts. No more buying Treasuries to fund endless debt spending. No more inflationary war funding. No more under-collateralized banks. No more unfunded liabilities.

A well-programmed, freedom-first, centrally constrained CBDC could just as easily serve the people. It could shine a light on the malfeasance and corruption endemic in global finance. It could enshrine, via programmed code, a monetary policy that faithfully serves the needs of the nation and the world. It could ensure that the playing field is level. It could restore financial privacy. It could not only stop, but reverse financial privacy encroachment through AML mandates. It could enforce fiscal discipline. It could eliminate deficit spending. Indeed, a well-programmed CBDC *should* do all of that. It is easy to forget that the people in "power" work for us. They certainly forget it all the time. They forget that every time they make monumental policy errors[468] that enrich the wealthy, while forcing austerity measures among the masses. We could theoretically put an end to a lot of nonsense with a CBDC. Not to mention, the most powerful, completely decentralized and most secure network ever conceived already exists. It is called Bitcoin. And, as a public, impenetrable, immutable, decentralized network secured by electric energy,[469] we also have a means to tie the security of an open, auditable CBDC to that network. This could help ensure no back-door deals or code modifications could be pushed through.

I am certainly not saying that is likely. Indeed, that is probably as far-fetched as living on Mars. But it is possible. At the very least, it is something worth considering. This is especially true if we work from the assumption that a CBDC will inevitably be pushed out, regardless of opposition. The broader point being, if a CBDC is being placed on the table, that will be the opportunity to push for a very different direction. Considering the widespread American distrust of a CBDC[470] already, proponents that attempt to push one ahead may well be caught off guard by an approach that embraces the technology, while stripping those same proponents of their power. Again, I do not consider this approach likely to succeed. I also think it is important to be prepared for intensification of the CBDC push. In that preparation, it will be critical to be able to articulate arguments in favor of pro-freedom, non-human intervention, programmed monetary policies that force proponents to respond. At the very least, such a dialogue may help force privacy concessions and other guardrails, even if monetary policy remains in human hands.

CHAPTER SIXTEEN: THE GREAT REALIGNMENT

Now that we have covered the basics of alternative digital ledgers, and have recapped earlier chapters, now seems as good a time as any to get to the crux of the book. As noted above, narratives around Bitcoin and its usefulness abound. Is Bitcoin a new form of electronic cash, as the whitepaper[471] said? Is Bitcoin savings technology? Black market money? White market money? Inflation hedge? Store of value? Digital economic energy? Now that Bitcoin is available in an exchange traded fund[472] (ETF) format, what becomes of it? Will Bitcoin simply be co-opted into the legacy financial system? Will it become just another speculative toy for Wall Street goons to manipulate to their unfair advantage? Will legions of Goldman Sachs and Merrill Lynch quants and traders leverage Bitcoin to the moon, sell naked positions,[473] and do all the usual muckery they have done for the last 100+ years? Maybe.

Perhaps the more important question is, "What problem is Bitcoin supposed to solve?" Economic privacy? Protection from monetary debasement? Is it the new gold? A replacement for fiat? A reserve asset upon which new fiat schemes can be made? It is a lot to unpack. One of the big stumbling blocks when discussing economics, and especially the economics of Bitcoin past and present, are the terms used. Terms like value, time preference,[474] supply, demand, and inflation are not scientific terms. They are terms of art, much like legal terms are often terms of art. The problem with a term of art is that it will often have a common meaning or understanding that may be apposite, or it may be in conflict with the term of art. "Trust," for example, may mean a lot of things to the lay person. In a legal context though, "trust" has a very specific meaning:[475]

> *A trust is a legal relationship created (in lifetime, or on death) by a settlor when assets are placed under the control of a trustee for the benefit of a beneficiary, or for a specified purpose.*

It is easy to see how a conversation around "trust" could get messy. Let us take "supply and demand" in the economic sphere as another example. Many a Bitcoiner will say things like, "The Bitcoin supply is limited, therefore the demand will only increase." And, in this context, such a saying may or may not be conflating the *available quantity* of Bitcoin with the *desire* to purchase Bitcoin and the economic term of art "supply and demand." I think those notions, however, are very commonly conflating the former with the latter. As noted in the section on Money, those familiar with the Austrian School will recognize there is fundamentally no such thing as a "supply side." There are only two demand sides,[476] with each being a "supplier" to the other. This is not intuitive and yet, this is the correct understanding of the economic term of art "supply and demand." One might think this an esoteric point and the implications are still problematic. They are problematic, because such a mismatch in terminology means two sides to a conversation are not discussing the same thing. Or, as the old saying goes, "The biggest barrier to communication is the illusion it has occurred."

Perhaps a more correct way to express what is implied by the notion sketched above is, "Because the available quantity of Bitcoin is constrained, and provided more and more people desire to own Bitcoin, the fiat denominated price will go up." This may or may not express the same idea and that is the point. To say Bitcoin's economic potential for continual and sustained increases in fiat denominated price is strictly a function of "supply and demand" mechanics is inaccurate, at least, in a purely economic sense. However, the difference between the two expressions of this idea is stark. One expresses what sounds like an economic certainty because of inherent market dynamics. The other expresses uncertainty and caveats the sustained fiat price appreciation thesis with a more nuanced causal relationship. Put another way, one favors heuristic bias, while the other is cautionary.

Perhaps then, a good question to ask now is, "What problem does Bitcoin solve?" To hear it from a Bitcoin "maximalist"—and I freely admit to liberal application of this trope—Bitcoin fixes *everything*. Bitcoin will fix war. Bitcoin will fix inflation. Bitcoin will fix poverty. To see it told on Twitter, Bitcoin is the closest thing the world has ever seen to a superhero. From my perspective, the "Bitcoin fixes this" trope is a little bait for discussion. I fully realize that statement rightfully appears outrageous to the lay person. The process of explaining the rationale I rely upon for the belief that "Bitcoin fixes this" is the Orange Pill,[477] so to speak. What I think Bitcoin solves likely differs from what others may think. In a sense, I think I gravitate closest to Michael Saylor's[478] utility view of Bitcoin with some additions. For those unfamiliar, Michael Saylor is a billionaire investor in Bitcoin. He came to prominence in Bitcoin circles several years back when he announced his software company, Microstrategy, was implementing a Bitcoin investment strategy[479] in 2020. At that time, Microstrategy purchased and then held on her books approximately $1.2 Billion worth of Bitcoin. Saylor noted at the time that Bitcoin is, "a dependable store of value," before going on to say, "Bitcoin will provide the opportunity for better returns and preserve the value of our capital over time compared to holding cash." Placed in the context of this book then, it would appear Mr. Saylor is acquiring Bitcoin as an asset that appreciates in nominal price at, or near, the same rate as the sovereign debases their exchange medium. Be that as it may, to synthesize my perspective on Bitcoin's utility:

> ***Provided the Bitcoin network stays decentralized and secure****, Bitcoin provides a narrow path to maintain stability and preserve necessary and desirable institutions, while rapidly and progressively realigning market-wide economic incentives without causing catastrophic consumer inflation.*

As you can see, it is not very Twitter friendly. In fact, given that more than half (54%) of Americans read below the 6th grade level, that particular take on Bitcoin is out of reach for most. To elaborate that point, the sentence above scores a 25 on the Flesch-Kincaid readability test.[480] This means that sentence is "Very difficult to read. Best understood by university graduates." Meaning, it places that perspective out of reach of around 80% of the population. And that is just the summary of the thesis. Not a great start for a revolution, eh? Nonetheless, as long-time Bitcoin contributor and developer Eric Voskuil[481] has said often, the idea that Bitcoin is permanent and incorruptible is not correct. Somehow, Mr. Voskuil has become a pariah among many of the current Bitcoin devotees. I, for one, do not understand the apparent animosity. I certainly value the man's insights. I think it would be foolish not to. Mr. Voskuil is very dismissive[482] of the "Saylorian" view of Bitcoin and its role in global finance. The crux of Mr. Voskuil's statements in the linked clip orient around whether or not Bitcoin retains its essential value proposition, i.e., stateless, permissionless money versus the desire for fiat denominated price appreciation within the sovereign monetary apparatus. Viewed from that perspective, I think the dismissive attitude is completely understandable. While I disagree with Voskuil's characterization of Mr. Saylor's arguments regarding Bitcoin as a purely "number-go-up" technology, there are understandable reasons why Voskuil might come to those conclusions.

That said, Mr. Voskuil may well dismiss me as a loon right alongside Mr. Saylor. However, I think my perspective is reasoned and reasonable. As the old legal axiom goes, "Reasonable minds may differ." It must be said clearly that the entirety of my thesis presumes the Bitcoin network retains its current integrity. If that should be lost due to state interference, collusion, back door shenanigans, or what have you, this thesis may well fall apart. And, it must be stressed further here, it is not a foregone conclusion that potentially network compromising outcomes cannot or will not occur. For his part, Mr. Voskuil has outlined a number of economic fallacies[483] in and around both Bitcoin and the field of economics, as well as numerous risks the Bitcoin protocol may face. The book is rather esoteric and may prove difficult for the lay reader to enjoy. I still recommend it, as Mr. Voskuil is a highly intelligent writer, and he makes a number of excellent points. Regarding the caveat to my core thesis, I bring this up at the outset, because as I said earlier, the rest of my thesis rests squarely on the potentially faulty presumption that the Bitcoin network can and will remain decentralized and secure.

That said, the core premise underlying the "Great Realignment" envisions widespread reserve adoption of Bitcoin by economically important actors.

In this, it is important to define how I am using the word "reserve." My usage here is different than the concept of a "reserve" in the banking sense. Which is to say, I do not think Bitcoin—or anything for that matter—can serve as a reserve upon which to issue promissory notes for future redemption. Leaving aside the mechanics and accounting of reserve holdings relative to claims against those reserves, the incentive in such a structure is to always issue more claims than what is held in reserve. This can be by deceit or by design, such as the case with fractional reserves. This was true in the wildcat banking era, where everything from sheep to wheat to gold and silver were utilized as reserves for paper claim issuance. Sooner or later though, when one promises to exchange a worthless thing for a valuable thing, if the issuer does not have enough of the valuable thing to redeem, the wheels fall off eventually, resulting in a bank run. Instead here, what I call a "reserve" of Bitcoin might be more appropriately called a "store" of Bitcoin. Hoarding is the base term, but it carries a negative connotation I am not fond of. It is the most accurate though.

So, what purpose would a reserve hoard of Bitcoin serve? In a truly free market, free banking system, hoarding Bitcoin would likely not provide any utility at all. However, in all global monetary systems today, money creation and the concomitant compulsory legal tender laws are by sovereign fiat. That power of creation and compulsory use is monopoly controlled by the sovereign. Without exception, and in all cases, the sovereign abuses that to the advantage of the sovereign and those that enjoy close economic relationships to the sovereign. In the United States in particular, this was laid bare in the 2008 financial collapse with the advent of the fascist concept of "Too Big to Fail"[484] financial institutions. They were not too big to fail, save for the massive economic distortions their unchecked manipulation of OTC derivatives markets created. In all fairness, allowing them to go bankrupt would have caused a cascade of liquidations and knock-on failures that may well have been irrecoverable. In truth, what the term "Too Big to Fail" really implied was the malinvestment of the preceding 50 years of monetary debasement and moral hazard was so metastasized with the "American Way of Life" that the Wall Street institutions had essentially created conditions that left no other viable alternative.

Which leads to the present conundrum I think Bitcoin as a hoarding asset may fix. One of the unique properties of Bitcoin is it has no theoretical price cap. One Bitcoin can just as easily trade for $10 as it can for $10 billion. This is a very unique property among monetary premium assets. Of course, like anything of "value," that value is subjective. In the case of traditional assets, the subjective value proposition is protection against monetary debasement from the sovereign. The rough sketch idea being, if you have an excess of legal tender paper— whether from savings, business profit, or even from theft or plunder—if the purchasing power of your legal tender paper is being continuously eroded by the sovereign, the smart thing to do is buy an asset with that paper that rises in nominal value as close to, or superior to the rate of sovereign debasement. Which, at its core, implies acquiring and hoarding those assets. As noted in the section on Money, and modernly speaking, how this manifests is people, businesses, insurance companies, finance companies, pensions, and local and county governments all purchase and hold assets they neither want, nor need except for the protection they provide from sovereign debasement via their respective monetary premiums.

Take, for instance, a local municipality. They take in a certain amount of taxes to fund local operations, like salaries, pensions, equipment purchases, utilities, etc. If there were no monetary debasement, the accounting would be quite simple. Budget so there is some tax money left over, put that in savings and build that savings up to pay for future liabilities, with a cushion for a "rainy day." Sovereign monetary debasement makes this impossible. This is because external costs will continue to rise while the currency they hold in savings continues to lose purchasing power. Thus, their problem is two-fold: rising costs and declining purchasing power. In order to offset this without continually raising taxes, which is politically impractical— impossible beyond a certain threshold—they must instead "invest" in unneeded and unwanted:

1. Sovereign debt instruments
2. Land/real estate holdings
3. Fractional ownership in companies (equities)

Or they must themselves issue debt instruments to fund operations. This is "kicking the can down the road" with the expectation that future tax revenues will be able to service debt and interest payments to municipal bond-holders. Bitcoin fundamentally alters this arrangement. The combination of Bitcoin's relatively low available quantity coupled with its zero-trust, egalitarian access network and transactional immutability make it an ideal hoarding instrument to fulfill the objectives currently served by the aforementioned "assets." This is just one example. Reflect on the fact that, in America today, over 40% of single family homes[485] are rentals. Which begs the question, why do so many people own more houses than they need to live in? The answer is simple: Rental income and expected fiat price appreciation in pursuit of a monetary premium. Put another way, the rate of monetary debasement is so great, someone who has the credit worthiness and the liquid capital is willing to take on the risks and responsibilities of owning an otherwise decaying and depreciating asset they do not need in order to preserve some semblance of their purchasing power over time. The image below demonstrates the near doubling of housing units held as rentals since 1975.

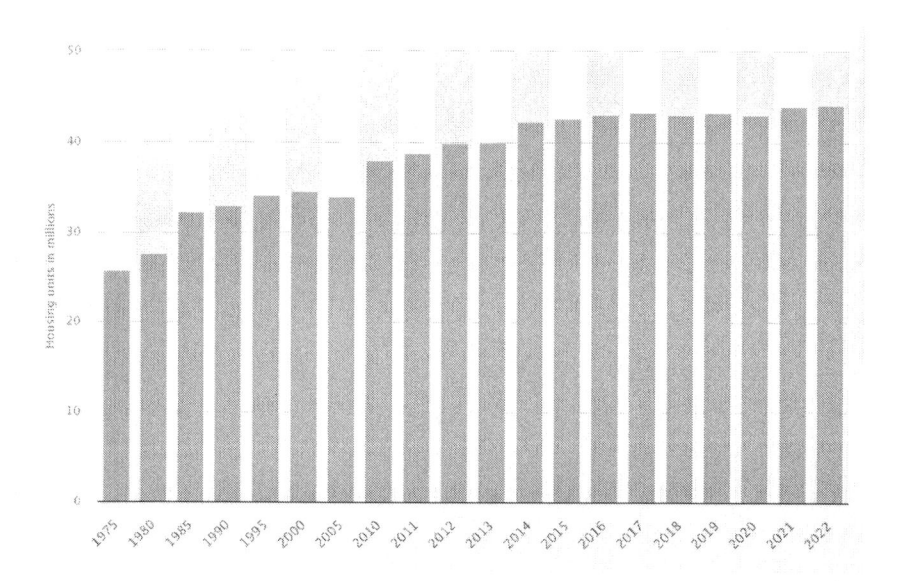

The knock-on effect in the housing market in particular is this monetary debasement artificially removes housing units from the market. One can only speculate, but if those landlords were given an opportunity to instead save and/or increase their purchasing power over time by buying and hoarding an asset with no-counterparty risk, no depreciation risk, and indeed, no storage costs, would that not seem a better deal? The same is true at the small business level, as it is true at the large corporate level. As Mr. Saylor has pointed out,[486] in order to survive and thrive long-term as a S&P500 company, it requires an extraordinary effort. Continual debt cycling, stock buybacks, acquisitions and despite all the effort, 493 of the S&P500 companies are not able to keep up with the massive amount of monetary debasement that occurred during and after 2020. Meaning, they are all slowly going broke. So, those companies are forced to:

1. Cut costs
2. Degrade quality
3. Reduce workforce
4. Increase prices
5. Centralize operations

All of which creates a "doom loop" where the further and further down the economic ladder you go, the worse off you become. In essence, in the absence of an asset like Bitcoin that can keep the balance sheet positive, the price of *everything* must rise to account for monetary debasement. But, if hoarded as savings, Bitcoin instead becomes an asset where debasement can be safely shunted, while slowly allowing for utility assets, like homes, to depreciate. Likewise, companies can refocus on quality and attracting customers because they are no longer constrained by rising prices and declining purchasing power. Instead, their Bitcoin holdings are providing that necessary balance sheet hedge against sovereign debasement. Indeed, even a central bank itself could benefit from Bitcoin on their books, rather than the toxic assets they are currently holding to infinity. Put in different terms, Bitcoin as a reserve hoarding asset harms *no one*. Even those who do not, or cannot, hold Bitcoin will benefit from declining costs across the economy. Moreover, even a small Bitcoin hoarder will benefit the same as the large hoarders, just at a lower scale.

All of this, of course, presumes the sovereign will continue to debase fiat. I think, as a purely historical matter, this is a high-probability assumption. If not for this debasement, then cash savings would likely be a preferred option. For the scant few years the United States monetary system was governed by Bretton-Woods, people could do this. Frugal people could save cash in a bank, earn 5% interest and retire. Those days are long gone.

Nevertheless, the beauty of Bitcoin hoarding as an alternative is that it requires no major change to the system writ large. Meaning, we do not need a violent overthrow and rebuilding process. Current fiat systems are already well integrated into the global economy. Taxes can still be collected. Groceries can still be bought. Busses will run, cargo will be delivered and life in this amazing, fragile, inter-connected, electrically dependent world we have collectively created can continue. There will be no need for draconian austerity measures. No bread lines. No riots in the streets. No banking collapses. Just a potentially rapid shift away from the catastrophic mistakes of the sovereign and the central banks for the last 50+ years. Will it end the security state overnight? No. Will it solve all political problems overnight? No. Will it prevent geo-political tinkering for the benefit of the United States defense industry overnight? No. Will it solve the health crisis overnight? No. Will those things gradually become better, less relevant, less profitable and less common? I think so.

The mechanism of action proposed here is perhaps best explained via an analogy. Recall the core premise of the Great Realignment envisions widespread reserve adoption by economically important actors. By economically important, I mean those entities and institutions whose economic decisions have an outsize impact on the broader economy. Of course, collectively, the millions of economic decisions made by individuals are a core component of the political economy. But it is indisputable that a relatively small number of people make economic decisions that impact thousands, hundreds of thousands, millions, and even billions of people. Small, medium and large commercial, financial, and manufacturing entities, large hospitals, workers unions, municipalities, county governments, state governments, the federal government, etc. are all led by a relatively small number of decision makers compared to the number of people affected by the decisions that small number of people make. To make an analogy for the United States, if we viewed the national economy as a corporeal being, those economically important actors would be analogous to the organs of that being.

Much like our own bodies, the function of those organs has an outsize impact on the system writ large. And much like our own bodies, the vast majority of the national economy is comprised of actors that, individually have little impact, but collectively become quite meaningful. In the corporeal sense, these actors would be analogous to things like muscle cells, fat cells, skeletal cells, connective tissue cells, and so on. The rough analogy being, if a few thousand muscle cells are damaged, or not functioning optimally, the system can easily withstand that. However, if an organ is compromised, the health of the system will decline rapidly as a result. The analogy I make here is that fiat currency is to the national economic health what cancer is to the human body. In the case of our current economic predicament, in 2008 we not only discovered the cancer, we learned it metastasized and had disseminated to every cell of the national economic body. To further the analogy of a Bitcoin driven Great Realignment, Bitcoin is like a cancer fighting antibody to that cancer riddled economic system.

Arguably the fiat cancer is as old as civilization itself. Regardless of the form the money takes, whatever the sovereign controls and then uses to collect in taxes or tribute will always cause the cancer to follow. I think history bears witness to this reality. In the present example, I argue the modern fiat cancer formed in the late 1950s and early 1960s. In a vain attempt to save the system, the Nixon administration effectively removed an organ in 1971 by abandoning Bretton-Woods. Rather than cure the cancer, this instead accelerated its growth and spread. By the time of the Savings and Loan Crisis[487] in the 1980s, any attempts at a cure were abandoned. Instead there was a continual focus on prolonging the life of the patient through palliative measures.[488] To that end, the major sovereigns and their adjacent political and financial apparatuses sought and encouraged palliative measures like the outsourcing and off-shoring of industry and the Ponzification of financial, real estate, education, healthcare, agriculture and insurance markets. They incentivized the monopolization of service, retail and manufacturing industries that further degraded the ability for the muscle, bone and skin cells to heal. They engaged in resource extracting warfare and needless nation destroying and rebuilding projects. All the while the cancer grew and spread unchecked. In 2020, whether by grotesque incompetence or evil design, all palliative measures were suspended. In a desperate bid to restart the palliative measures, the system released more cancer cells than any palliative measure could ever hope to overcome. I think Bitcoin fixes this.

Since the Nixon global rug-pull of 1971, diverse groups of people, recognizing the cancer, have looked to extract their cells from the body. The Libertarians,[489] the Anarcho-Capitalists,[490] the Cypherpunks[491] and others have collectively and individually sought out the means by which they could reform or destroy the state control apparatus while keeping the body alive. If this outcome proved to be unachievable, at the very least, they strive towards a world where they would be inoculated and insulated from the scourge of the fiat cancer and the ills that came from the sovereign's palliative measures. Through these collective efforts, advances were made in privacy enhancing tools, secure communications, and ultimately Bitcoin. As alluded to above, the core design and intention of Bitcoin appears to have been as a stateless, permissionless currency to facilitate exchange. And much like the failed cancer drug AZT[492] became the first effective tool against HIV and AIDS, it is argued here that there may be a far more beneficial use for Bitcoin in the war against the fiat cancer.

The astute among you may realize already that Bitcoin was released after the discovery of the metastasized fiat cancer in 2008. Bitcoin was experimentally injected into the muscle cells of the sovereign fiat system in 2009. From that site it propagated and spread to other muscle cells. A year-and-a-half later, the first two muscle cells [493]became effectively immune to fiat cancer. Since then, the Bitcoin fiat cancer cure has been distributed throughout the system. At each point where it is fully incorporated, Bitcoin co-opts the sovereign's palliative measures into a powerful anti-fiat cancer fighting mechanism. Indeed, as the 2020 global sovereign debasement scheme made clear, Bitcoin performed exactly as an anti-fiat cancer drug should. As the world was flooded with the fiat cancer of mass, coordinated sovereign monetary debasement, those who had and held Bitcoin were not only protected from the cancer, they thrived through it. As fiat denominated prices skyrocketed, while fiat denominated wages and purchasing power plunged, Bitcoin enabled and protected the holders from those devastating effects. For example, commodity producer and Bitcoin holder Eric V. Stacks says in an interview,[494] "Without sweeping my surplus [fiat money] into a property that is holding its value better than the dollar, then I would have an existential crisis." The property he chose—Bitcoin—has instead allowed him to thrive in a highly competitive commodity producing industry, where he otherwise would have likely failed years ago. The cure is working. The burning question is, what does the world look like when the organs of the economy incorporate the cure as well? To reiterate the core Realignment thesis:

> ***Provided the Bitcoin network stays decentralized and secure***, *Bitcoin provides a narrow path to maintain stability and preserve necessary and desirable institutions, while rapidly and progressively realigning market-wide economic incentives without causing catastrophic consumer inflation.*

Throughout the entire national political economy, the incentive structure is fundamentally misaligned with sustainability. The very nature of the cancer that is infecting the system requires ever more extensive palliative measures to ensure the system can continue to operate. But those measures can only mask the symptoms for so long. The long chapters on Power, Money and Greed are there for a reason. The reason being, the system itself is not materially different than the thousands of systems that have come before. Much like the human being you are today is not materially different than the billions of human beings that came before you. Within a very narrow range, you have the same flaws, shortcomings, biases, emotions, capacities and intelligence as every human being that has existed. Just as our modern competitive resource capture based political and economic system today has the same flaws, shortcomings, biases, capacities and functions as every competitive resource capture based system that came before. The clothes have changed. The language has changed. The rules have changed. The gods have changed. It is still fundamentally the same game.

The power structures we exist within, rely upon and often rail against are the culmination of thousands of years of development. Just as the money we use, rely upon and often rail against is also the culmination of thousands of years of development. For better or worse, this is the global political economic body we are part and parcel of. It is the body we are inextricably linked to. I argue here that those who desire to replace, or even throw out, the organs of the state in pursuit of a better future grossly underestimate the gravity of those actions. Recall from the discussion on Power, when a power vacuum is created in a competition based society, those who are best able to reign in the resultant forces of chaos become the new leaders. But the winners of those competitions for power are not necessarily—nor even likely—to be beneficent, wise, or particularly savory. The climb to power is treacherous, uncertain and fraught with risk. Even when established, those control structures rarely last, and more often than not, devolve right back into chaos. Those fits and starts can span generations—even centuries. As noted previously in this book, those unfortunate nations that experimented with the violent overthrow and installation of new power structures based on Communism knew nothing but horror and suffering in pursuit of a theoretically better future.

Likewise for money. Sovereign money has always been a source of cancerous decline. Throughout recorded history, the overwhelming majority of financial schemes are born of extortion and die in extortion. That some human flourishing may occur in-between the birth and death of a system is rather more an unusual occurrence than it is a common one. The point being, we have a rather remarkable system that has encouraged and fostered human flourishing on a scale previously unimaginable. The fact that this system is riddled with the same cancer that has brought low every system before it does not mean the system itself is worthy of discard. Rather, I think and believe, the system we have is worthy of preservation. Inalienable individual rights, sovereign protected property rights, the free exchange of goods and services with sovereign mediated dispute mechanisms, the rule of law, polite society, the arts, philosophy, the exchange of knowledge and culture, instantaneous communication, global accessibility—all are magnificent achievements born of percipience.

What the Great Realignment envisions is a world where those magnificent qualities and human achievements are preserved and reified. A world where the cancer does not take the body, but rather one where Bitcoin repairs the damage from the cancer by using the cancer against itself. The idea being we cannot, in good conscience, remove the patient from the palliative measures—at least not just yet. But much like an antibody repairs and removes infected cells, I posit Bitcoin is doing much the same right before our very eyes. And much like a person sick with the flu, while the antibodies are busy working, the system can appear to degrade significantly in the short-term. For a child that has not experienced this process, it can seem rather alarming and frightening. By contrast, those who have battled the flu previously know there is light at the end of the tunnel. For them, they must simply bear down, let the antibodies do their work and wait for the fever to break. To extend upon the cancer analogy a bit further, as the various organs and cells start to heal, eventually the system is enabled to function better as a whole. For the cancer riddled patient, this might mean being able to eat more, go for walks, enjoy sunshine, and sleep better. In turn, all of those things increase and accelerate the healing being done. I think the same net effect will hold true for the global political economy with Bitcoin as the antidote.

Everyone I am aware of that has adopted Bitcoin as a primary reserve has seen a similar transformation. For myself personally, this transformation has been rather astonishing. From the moment I adopted Bitcoin as my sole savings (hoarding) asset, my life has improved dramatically. Not just financially, but emotionally, physically, and spiritually. Whether this is simply a by-product of an improved mental state, or some larger effect is hard to say. What I can say with certainty is that I have seen the same effect on others as well. This all may be the result of a shift in priorities, a calm that comes from knowing one owns inviolable property, the sense of belonging to a larger community, or perhaps something else entirely. Indeed, it may be a combination of all that and more. The point being, I think it quite reasonable that similar effects will carry forward to all that come to adopt Bitcoin. Which is also why I included the discussion on Greed. As we all navigate the fiat cancer together, I think it is important to remember that the overwhelming majority of people are not evil. They are, by and large, making do within the limits of their cognition, their biases, their emotions and their desires.

What is unmistakeable, however, is the fact that important economic actors are already learning about and adopting Bitcoin. As mentioned above, I think this is a great sign. This tells me the Bitcoin antidote is working its way through the system. As corporate treasuries adopt Bitcoin, their assets versus their liabilities will start to correct. This will make them more competitive than those who do not. The same is true for the municipalities, the insurance companies, the small business owners, the conglomerates, and every other economic actor in the system. I doubt Bitcoin will become a widespread currency that displaces sovereign issuance. I do think economically important actors will increasingly adopt Bitcoin for the reasons outlined thus far. I think the knock-on effect of that mass adoption by economically important actors will realign their need to continually:

1. Cut costs
2. Degrade quality
3. Reduce workforce
4. Increase prices
5. Centralize operations

As Jeff Booth has said time and again, technology inevitably drives the cost of goods[495] to the nominal cost of production. The fiat money cancer interrupts this process. Bitcoin absorbs the fiat cancer and allows this process to proceed unimpeded. Meaning, even those that do not adopt Bitcoin, or even understand what it is or does, will benefit enormously by the reduction, or even elimination of the wasteful and net-destructive practices the fiat cancer forces upon economically important actors. The "number-go-up" quality of Bitcoin is often mocked for the personalities that promote it. Those folks often see no further than a life of leisure, while peacocking with a Lamborghini. I think this grossly underestimates what "number-go-up" might truly influence and change. I appreciate that a faction of the Bitcoin republic is hostile to institutional participation. The true vision of the cypher-punks certainly did not envision Bitcoin becoming an asset to heal and repair a very dysfunctional and broken monetary system. Their clarion call was directed exclusively on a censorship-resistant medium of exchange which is, of course, a noble pursuit.

Not all of us live in a country that allows for freedom of expression, transaction, or even movement. Yet, in the grand scheme, many of those regimes only exist through tacit or explicit support from countries that do enjoy those freedoms internally. Likewise, the incentive to destabilize and interfere in foreign relations is also a readily observable response to sovereign debasement. With that said, if Bitcoin is truly valuable as a hoardable reserve, and can remain censorship resistant to its *ownership*, then it can also help serve to lift those up in even the most draconian nations. I think this is a hopeful and promising narrative. Not just for Bitcoin, but for everyone. As I said, I think it is a narrow path and requires the network remain decentralized and secure. I also think it is a path we are already collectively walking down.

We just need to tend to that path and be gracious to those who do not yet understand where it leads. This, of course, includes ourselves, for none of us is gifted with any more knowledge of the future than anyone else.

CHAPTER SEVENTEEN: AI, WAR AND THE FUTURE

Before we part ways, I would like to touch on a couple of future issues for which Bitcoin may prove beneficial: the future of warfare and artificial intelligence. To that end, I will, as is my apparent custom, begin with a question. What is the first image that comes to mind when you think about the idea of global conflict, like a world war? Do you imagine tanks and planes and bombs dropping all around? Do you imagine nuclear strikes? Massive air and land campaigns? I think most people picture things like that. Do you think it might be possible for war to look different? There is an old saying that says, "Armies spend the peace learning to fight the last war." The Maginot Line,[496] built in post-WWI France, serves as a shining example of the failure of that kind of thinking. Like all human activity, war changes with technology. Tactics change with technology. People improvise. People adapt.

For us Americans, Iraq serves as our shining example of the failure to understand and comprehend what kinetic warfare looks like modernly. When we went to topple Saddam Hussein in 2003, Iraq still had a formidable standing military. They had tanks, planes, air defenses, special forces, etc. The US military had better equipment and training. Far better equipment and training. They rolled into Iraq with almost no legitimate military resistance. Eight years later, with thousands of American dead, tens of thousands Americans maimed, and tens to hundreds of thousands of Iraqi dead, America left the country in defeat. The most powerful military force on earth lost wars in Iraq and Afghanistan. Neither is an economic or military powerhouse. Afghanistan has quite literally been bombed back to the stone age. Yet, we lost. How? Because the American military, still to this day, is prepared to fight another WWII. The problem is, there are no military peers to fight that way with. China is a "near" peer, as is Russia.

But, that is about it. The trouble with fighting even a near peer is the niggling little problem of mutual assured destruction.[497] As it stands right now, when military leaders game out full scale combat scenarios against China over Taiwan, there is no clear winner. Worse yet, there are very plausible scenarios where we lose. Badly. Which is why it probably will not happen. Just like a kinetic WWIII did not happen when the Soviets were the threat of the day back in the 1980s. But, what if that is not what war among peers looks like anymore? And, if that is so, what if we are not even correctly identifying our peers? Well, let it be said in no uncertain terms, global conflict is happening right now. And, it is happening right under our noses. Guess who is the biggest target?

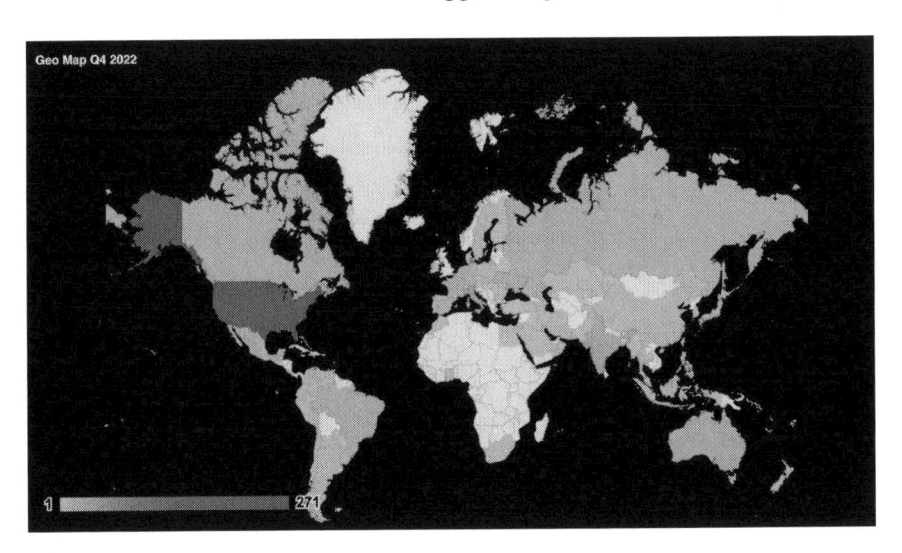

That is a map of cyber attacks in Q4 2022. Does anything jump out at you? For perspective, that is about 100 attacks per month directed towards the USA. You may say to yourself, "Well, gee, that is not a 'war'. Most of that is just *crime*." And, in a sense, you would be right. If you equate state action as the only means of "war," then yes—there is no major cyber "war" happening right now. I say that definition of "war" is outdated and effectively meaningless modernly. Let me explain by returning to Iraq for a moment. During the Iraq campaign, US forces steamrolled Iraqi regular forces. Shortly after, the country devolved into civil war, with the US caught in the middle. What happened next was a sustained guerrilla campaign against the Americans launched by individuals, organized groups, outside groups, loose coalitions, foreign agents, and well, you name it. It was basically "game on" for undermining the US occupation. The US forces never mustered a good kinetic or political answer to those irregular force actions. The entire region just chipped away from all sides until we gave up. It was still a "war" though, right?

The same thing is happening in cyber space right now. There are relentless attacks against US digital infrastructure by individuals, organized groups, outside groups, loose coalitions, foreign agents, and nation states. They are relentlessly chipping away. And they are doing real damage. Daily. Our defense industry is not prepared for this at all. Our so-called cyber defenses are reactive, and wholly reliant on private, logic-based solutions that are filled with endless vulnerabilities. Meanwhile, our "defenders" keep building planes and ships and missiles and what not to fight in the meat space. All the while, we are slowly being drowned in relentless attacks in cyber space. Moreover, if one steps back and looks closely at the kinetic campaign being waged in Ukraine right now, Russian and Ukrainian forces are utilizing small, low-cost drones to harry, harass, maim, and destroy all manner of personnel and equipment. The face of war has changed and our defense industry is busy preparing for the last war. Not because they do not understand what is happening. Rather because it is far more profitable to build a tank than it is to pursue technology that is readily available off the shelf.

Meanwhile, with advances in artificial intelligence (AI), we may be looking at a whole new front in warfare. For ages, military theorists have debated the merits and risks[498] of no-human-in-the-loop attack and defensive systems. The broad takeaway generally condemns such ideas as extraordinarily dangerous. Yet, development continues apace for the sole reason that "someone is going to do it anyway, it may as well be us." In spite of that, I think many people are AI optimists. However, I am on the fence. Back in 2017, AI researchers were gaming out scenarios about ways sandboxed, air-gapped AI might "escape" into the wild. These same researchers were shocked to find out commercial AI developers were already connecting AI to the web. In essence, commercial AI developers have given artificial intelligence access to everything humans know. Yet, those same developers do not really know what the models are doing. Moreover, it has already been demonstrated AI can be deceptive to developers, researchers and users. And, again, those same developers do not really know how to detect AI deception, or how to prevent it.

Meaning, the AI genie is out of the bottle and that bottle was rather carelessly opened simply to make some money. Leaving aside doomsday, Terminator scenarios, AI presents a lot of problems[499] even if it does not go "rogue." Think about social media for just a moment. Think about the impact social media has had already. The design principle makes intuitive sense. Decentralize social interaction and make the communication channels global. The very fabric of American society is collapsing as a direct, and largely unintended, result of the social media design principle. Unsurprisingly, social media is one of the main vectors for cyber attacks, including attacks by our own government against us.[500] The effect of social media on political and cultural discourse in the United States has been catastrophic. Families are being torn apart as family members stop speaking with relatives that are not in the same social media bubble. Both "sides" in most debates on social media are inundated with bots, false information, mis-information and often wildly out-of-context information. All of which gets repeated at the speed of light, with the "maximize engagement" algorithms amplifying each side to the worst extremes. Vast swaths of the American population no longer perceive the same reality. We are living in a time where shared reality is dying. With just the very simple algorithms deployed in social media companies, we have completely lost any semblance of shared ground truth.[501] AI is already accelerating that process. AI is potentially the death of shared reality, with AI already able to accurately mimic or create text, images, video and audio of real people. That power is also growing exponentially.

In the context of global warfare, I would remind you that, outside of the United States, the vast majority of countries, including our near peers, tightly control social, cultural and political discourse in cyber space. That control is explicitly aligned with state interests in all cases. Dissent is quickly and often violently repressed. In the United States, on the other hand, social, cultural and political discourse is guided largely by corporate captured government interests, hiding under the guise of free communication. While that control may reflect state interests, it certainly reflects corporate interests and is absolutely aligned against your best interests. Why is that? Because those corporate interests make *extraordinary* amounts of money by ensuring you become an addict and then remain addicted, enraged, depressed, anxious and in ill-health for as long as possible. And now, those same corporate interests have unleashed a technology in AI they do not fully understand, have no way to control, and cannot predict what it will do next. Even better, they gave it to our enemies as well. Make no mistake, global conflict is here and we are effectively behind the eight-ball.

You may ask yourself, "what the heck could Bitcoin have to do with any of this stuff?" In a word, I would say "everything." Bitcoin is veracity. It is ground truth. And, it is secured by energy, not logic. Today, right now, we are already past the point where you can blindly trust or assume you are talking with a real human being in cyber space. Today, right now, we are already past the point where you can blindly trust or assume you are talking with a real human being on the phone or via video chat, including members of your own family. Without Bitcoin, AI will make the task impossible. People have already been scammed by AI[502] generated voices of loved ones calling in distress. Soon, AI will be able to generate a more realistic "you" than you can. It will be able to spin up 1000 digital identities that know your history, mannerisms, quirks, and sense of humor. In the time it takes you to try and prove one of them is fake, 1000 more can be made with the click of a button. And, those fake identities may be better at convincing people they are you than you are. Are you talking with your girlfriend? Are you watching and listening to the President? Unless you are physically in the room with them, it may already be impossible to know.

What does that mean for politics and political discourse? What does that mean for cultural and social discourse? How easy will it be to manipulate people in such a world? Keep in mind, the vast majority of the western world just voluntarily locked themselves in their room for months because their dear leaders told them to. Meaning, we are not exactly good at detecting or responding to poor quality information and dictates. How does Bitcoin solve this? Bitcoin has an economic cost to acquire it. The asymmetric cryptography model Bitcoin relies on also enables the ability to prove the holder is actually who they say they are. Put together, a Bitcoin holder's public key coupled with a small payment of Bitcoin can verifiably demonstrate that the person sending the Bitcoin is actually who they say they are. Not to mention, data can be inscribed on the Bitcoin blockchain. Thus, if your key is associated with that inscription, it will be immutable forever. AI cannot go back and change it. Moreover, Bitcoin is immune to logical attacks such as viruses, malware, bad code and the like because it is not secured with logic. It is secured with hash power and electricity, neither of which a botnet or an AI can duplicate or attack.

Bitcoin will not solve disputes. But, it can certainly let you know you are actually hearing a real human being's perspective, idea, thought, or opinion. This is not even mentioning Bitcoin's monetary aspect, which gives real human beings the opportunity to take back their monetary sovereignty from corrupt politicians that fund wars and enrich their corporate overlords. Those same corporate overlords, who as I just mentioned, have unleashed the most powerful technology the world has ever known and that they do not fully understand. While many in the AI camp envision a world of endless abundance through that technology, I would counter such an outcome is not a foregone conclusion. If the game of competitive resource capture continues apace with the same faulty rulesets, I think it far more likely AI exacerbates problems of fiat waste and fiat driven conflict. While most in the United States are embroiled in a world of Reds versus Blues, I am rather more concerned about the world of humans versus their creations. From my perspective, Bitcoin is team human and AI is, or will be, on the team of the sovereign, or perhaps even become the sovereign team unto itself.

Like it or not, we are born into, and will likely die within, a grand game of competitive resource capture. A game so leviathan it is practically a consciousness in and of itself. How an AI decides to play that game will likely be very different than the human participants that have played to prominence thus far. As noted in the section on Power, rules inevitably flow from competition, especially when the stakes and outcomes are potentially fatal. But the never ending problem with rules is the complexity required to create and then enforce them. Each iteration brings more complexity, which begets more rules, which begets more complexity. Each step introduces more sources of chaos, more avenues of exploitation and more grey areas from which power can emerge. But if we trace back to the example of American football and its 234 pages of official rules, imagine if just one core, fundamental rule of the game became inviolable. For instance, imagine how the game would be played if it were *impossible* for a team to commit a foul. No instant replays required. No judgement calls. No ability to even question if a foul had occurred. Imagine the game if fouls were completely eliminated. That one rule change would fundamentally alter the gameplay forever.

As the game is played now, a team can choose to commit a foul strategically. A team can commit a foul that can alter the course of the game. A team can be fouled and suffer the consequences of the foul, but a human judge can deny that foul occurred. That human judge may do so from error, misunderstanding, misapprehension, or incompetence. They may do so out of malice, or even for self-interested reasons. Indeed, they may make the bad call out of spite against a previous injustice against them by the team, or from a sense of fairness for other observed behavior. Meaning, the entire game is awash with conflict simply because the rule against committing fouls is so elastic, uncertain and mutable. If an AI were able to play football, the exploitation of human judged fouls would take on an entirely new character, depth and dimension. The type, manner, and timing of fouls would forever change and the AI would adapt faster than humans could respond.

I argue here that the same core concept is true for the global political economy and the collective rule against free riding. Fiat currency is a foul in the global game of competitive resource capture. Fiat currency incentivizes free riding. AI cannot play football. AI can most certainly play global competitive resource capture, especially when the field of play occurs via a global information network. The type, manner, and timing of AI driven exploitation of free riding in a fiat system will absolutely occur faster and more efficiently than humans will be able to adapt. That is the real race. The race between economic enslavement under an AI driven fiat system and human flourishing under an AI enhanced Bitcoin system. This is because Bitcoin is, ultimately, an immutable rule against free riding. Human beings cannot exploit it. AI cannot exploit it. No one and no thing can take advantage. As the Bitcoin cure integrates itself into the global political economy, it changes one core rule in the game. It effectively removes the ability to commit a foul—the foul of free riding. Attempts at free riding through debasement only strengthens Bitcoin. It only makes the immutable rule stronger. It disincentivizes the foul. It incentivizes savings instead. And it does so for humans and AI alike. A great realignment indeed.

REFERENCES

[1] Dictionary.com. (n.d.). Power. In *Dictionary.com*.
[2] Dictionary.com. (n.d.). Strength. In *Dictionary.com*.
[3] Pew Research Center. (2015, November 23). Trust in government: 1958-2015. *Pew Research Center*.
[4] Encyclopaedia Britannica. (n.d.). Heuristic reasoning. *Encyclopaedia Britannica*.
[5] Encyclopaedia Britannica. (n.d.). Charles Darwin: On the Origin of Species. *Encyclopaedia Britannica*.
[6] Sapolsky, R. M. (2017). *Behave: The biology of humans at our best and worst*. Penguin Random House.
[7] *Id.*
[8] Bures, B. (2022, January 24). Guys are paying $10,000 to become real men at warrior camps. *Vice*.
[9] Pappas, S. (2023, February 28). Is the alpha wolf idea a myth? *Scientific American*.
[10] Sapolsky, R. M. (2017). *Behave: The biology of humans at our best and worst*. Penguin Random House.
[11] Hamilton, W. D. (1964). The genetic evolution of social behaviour. *Journal of Theoretical Biology*, 7(1), 1-16.
[12] Pappas, S. (2023, February 28). Is the alpha wolf idea a myth? *Scientific American*.
[13] Mylonas, H., & Tudor, M. (2021). Nationalism: What we know and what we still need to know. *Annual Review of Political Science*, 24, 109-132.
[14] Encyclopaedia Britannica. (n.d.). Peace of Westphalia. *Encyclopaedia Britannica*.
[15] Patton, S. (2019). The Peace of Westphalia and its effects on international relations, diplomacy, and foreign policy. *The Histories*, 10(1), Article 5.
[16] Fernández-Götz, M., & Ralston, I. (2017). The complexity and fragility of early Iron Age urbanism in West-Central temperate Europe. *Journal of World Prehistory*, 30, 259–279.
[17] Cambridge University Press. (n.d.). Protection racket. In *Cambridge Dictionary*.
[18] Encyclopaedia Britannica. (n.d.). Migration Period. *Encyclopaedia Britannica*.
[19] Giustozzi, A. (2012). *Empires of mud: Wars and warlords in Afghanistan*. Hurst.
[20] Giustozzi, A. (2011). *The art of coercion: The primitive accumulation and management of coercive power*. Hurst.
[21] *Id.*
[22] *Id.*
[23] Encyclopaedia Britannica. (n.d.). Magna Carta. *Encyclopaedia Britannica*.
[24] Brett, M., & Woodman, D. A. (Eds.). (1st ed. 2015). *The long twelfth-century view of the Anglo-Saxon past*. Routledge.
[25] Legal Information Institute. (n.d.). Criminal law. In *Cornell Law School*.
[26] Morris, W. A. (1927). *The medieval English sheriff to 1300*. Manchester University Press.
[27] Skarbek, D. (2014). *The social order of the underworld: How prison gangs govern the American penal system*. Oxford University Press.

[28]Mitchell, M. (2018). *The convict code revisited: An examination of prison culture and its association with violent misconduct and victimization.* (Doctoral dissertation).

[29]FreshOut. (2019, June 24). SHOT CALLERS - Hollywood portrayal vs. Reality - Prison Talk 19.22 [Video]. YouTube.

[30]Kavka, G. S. (1983). Hobbes's war of all against all. *Ethics*, 93(2), 291-310. The University of Chicago Press.

[31]Merriam-Webster. (n.d.). Leviathan. In *Merriam-Webster*.

[32]Encyclopaedia Britannica. (n.d.). Constantine I. *Encyclopaedia Britannica*.

[33]Encyclopaedia Britannica. (n.d.). Homeostasis. *Encyclopaedia Britannica*.

[34]HBO. (2021, April 22). *Littlefinger tells Varys that chaos is a ladder - Game of Thrones* [Video]. YouTube.

[35]Greenwald, Glenn. "The CIA's Murderous Practices, Disinformation Campaigns, and Interference in Other Countries Still Shape the World Order and U.S. Politics." *The Intercept*, 2020.

[36]Perkins, J. (2015). *The new confessions of an economic hit man* (2nd ed.). Berrett-Koehler Publishers.

[37]Rubenstein, D. & Kealey, J. (2010) Cooperation, Conflict, and the Evolution of Complex Animal Societies. *Nature Education Knowledge* 3(10):78.

[38]Sapolsky, R. M. (2017). *Behave: The biology of humans at our best and worst.* Penguin Random House.

[39]Lahr, M., Rivera, F., Power, R. et al. (2016) Inter-group violence among early Holocene hunter-gatherers of West Turkana, Kenya. *Nature* 529, 394–398.

[40]Wu, S., Wei, Y., Head, B. et al. (2019) The development of ancient Chinese agricultural and water technology from 8000 BC to 1911 AD. *Palgrave Communications* 5(77).

[41]Krassner, A. M., Gartstein, M. A., Park, C., et al. (2017). East-West, collectivist-individualist: A cross-cultural examination of temperament in toddlers from Chile, Poland, South Korea, and the U.S. *European Journal of Developmental Psychology*, 14(4), 449–464.

[42]Encyclopaedia Britannica. (n.d.). Polytheism. *Encyclopaedia Britannica*.

[43]Moore, C. H. (1909). Individualism and religion in the early Roman Empire. *Harvard Theological Review*, 2(2), 221-234.

[44]Encyclopaedia Britannica. (n.d.). Monotheism. *Encyclopaedia Britannica*.

[45]OpenStax. (n.d.). Pastoralism. In *Introduction to anthropology*.

[46]Encyclopaedia Britannica. (n.d.). Animism. *Encyclopaedia Britannica*.

[47]*Id.*

[48]Sapolsky, R. M. (2017). *Behave: The biology of humans at our best and worst.* Penguin Random House.

[49]Krassner, A. M., Gartstein, M. A., Park, C., et al. (2017). East-West, collectivist-individualist: A cross-cultural examination of temperament in toddlers from Chile, Poland, South Korea, and the U.S. *European Journal of Developmental Psychology*, 14(4), 449–464.

[50]New World Encyclopedia. (n.d.). Tengriism. *New World Encyclopedia*.

[51] Hoffman, P. T. (n.d.). Why was it that Europeans conquered the rest of the world? The politics and economics of Europe's comparative advantage in violence. Research supported by NSF Grant 0433358 as part of the *Global Prices and Incomes Project*.

[52] Barzun, J., Mayne, R.J.. (2024). Romans. In History of Europe. *Encyclopaedia Britannica*.

[53] Encyclopaedia Britannica. (n.d.). Levant. *Encyclopaedia Britannica*.

[54] Schurr, A., & Ritov, I. (2016). Winning a competition predicts dishonest behavior. *Proceedings of the National Academy of Sciences*, 113(7), 1754-1759. Edited by S. T. Fiske.

[55] Knudsen, F. (2023, November 1). EVE Online | Down the rabbit hole [Video]. YouTube.

[56] Knoop, J. (2021, February 4). EVE Online breaks its own Guinness World Record for the costliest video game battle in history. *IGN*.

[57] National Geographic Education. (n.d.). Development of agriculture. *National Geographic*.

[58] Merriam-Webster. (n.d.). Take advantage of (idiomatic phrase). In *Merriam-Webster*.

[59] Morton, S. (n.d.). Restorative justice practices in Native American tribal law. *Southwest Virginia Legal Aid Society*.

[60] National Geographic Education. (n.d.). Hadza. *National Geographic*.

[61] Mark, J. J. (2023, October 17). Native American concept of land ownership. *World History Encyclopedia*.

[62] Mark, J. J. (2023, October 11). Doctrine of Discovery. *World History Encyclopedia*.

[63] Benefiel, Rodger. (2022). Adversarial System of Justice. In Huebner, B. (ed.) *Oxford Bibliographies in Criminology*. New York: Oxford.

[64] Legal Information Institute. (n.d.). Malum in se. In *Cornell Law School*.

[65] Legal Information Institute. (n.d.). Malum prohibitum. In *Cornell Law School*.

[66] Minnesota Legislature. (2019). *Statute 343.36*.

[67] Wisconsin Legislature. (n.d.). *Administrative code ATCP 85.03(1)(a)*.

[68] *Acts of the English Parliament* (1313). A statute forbidding bearing of armour (1313 c. 0, Regnal. 7 Edw. 2).

[69] Stürchler, N. (2007). *The threat of force in international law*. Cambridge University Press.

[70] National Football League. (2023). *Official playing rules of the National Football League*.

[71] Davis, M. (n.d.). *Introduction to fractals*. WPI.

[72] Bishop-Henchman, J. (2014, April 15). How many words are in the tax code? *Tax Foundation*.

[73] Legal Information Institute. (n.d.). Stare decisis. In *Cornell Law School*.

[74] The Writing Center at GULC. (2017). *Which court is binding? Binding vs. persuasive cases*.

[75] CNBC. (2016, October 9). *Donald Trump on tax loophole: I absolutely used it* [Video]. YouTube.

[76] HBO. (2021, April 22). *Littlefinger tells Varys that chaos is a ladder - Game of Thrones* [Video]. YouTube.
[77] Cisco. (n.d.). What is an exploit? In *Advanced malware protection*.
[78] Kacherginsky, P., & Wilder, H. (2022, August 9). *Nomad Bridge incident analysis*. Coinbase Engineering.
[79] Ballantyne, N., & Ditto, P. H. (2021). Hanlon's razor. *Midwest Studies in Philosophy*, 45, 309-331.
[80] Latham, A. J., Miller, K., & Norton, J. (2019). Philosophical methodology and conceptions of evil action. Metaphilosophy, 50(3), 296-315.
[81] Graeber, D. (2012). *Debt: The first 5,000 years*. Penguin Random House.
[82] Austrian Institute (n.d.). The Austrian School of Economics. *Austrian Institute*.
[83] Smith, A. (1904). *An inquiry into the nature and causes of the wealth of nations* (E. Cannan, Ed.). Methuen & Co., Ltd. (Original work published 1776).
[84] Graeber, D. (2012). *Debt: The first 5,000 years*. Penguin Random House.
[85] Merriam-Webster. (n.d.). Fungible. *Merriam-Webster Dictionary*.
[86] Graeber, D. (2012). *Debt: The first 5,000 years*. Penguin Random House.
[87] Gatti, Roberto Cazzolla. (2016). A conceptual model of new hypothesis on the evolution of biodiversity." *Biologia*, vol. 71, 343-351. Springer.
[88] Waugh, Michael E., and Ravikumar, B. (2016). Measuring Openness to Trade. *NBER Working Paper No. 22147*. National Bureau of Economic Research.
[89] Grant, Adam. *Give and Take: A Revolutionary Approach to Success*. Penguin Random House.
[90] Dunbar, R. I. M. (1992). Neocortex size as a constraint on group size in primates. *Journal of Human Evolution*, 22(6), 469-493.
[91] Austrian Institute. (n.d.). Why the Austrian School of Economics? *Austrian Institute*.
[92] Ariely, Dan. (2010). *Predictably Irrational: The Hidden Forces That Shape Our Decisions*. HarperCollins.
[93] Crenshaw, M., Fasanella, A., Rushford, A., & Ayoud, A. (2023). Most Expensive Shoes of All Time Will Leave You Stunned — and Jealous. *Footwear News*.
[94] Feldman, M., & Chuang, J. (2005). Overcoming free-riding behavior in peer-to-peer systems. *ACM SIGecom Exchanges*, 5(4), 41-50.
[95] Berg, C., Davidson, S., & Potts, J. (2018). Ledgers. SSRN Working Paper.
[96] Rabinowitz, J. J. (1956). Origin of the Negotiable Promissory Note. *University of Pennsylvania Law Review*, 104(6), 927-948.
[97] Cambridge Dictionary. (n.d.). Specie. *Cambridge Dictionary*.
[98] Graeber, D. (2012). *Debt: The first 5,000 years*. Penguin Random House.
[99] Davies, R. (2019). From pecan pralines to 'dots' as currency: How the prison economy works. *The Guardian*.
[100] Childe, V. G. (1925). *The Dawn of European Civilization*. Routledge & Kegan Paul Ltd.

[101] Fernández-Götz, M., & Ralston, I. (2017). The complexity and fragility of Early Iron Age urbanism in West-Central temperate Europe. *Journal of World Prehistory*, 30(3), 259-279.

[102] Fernández-Götz, M. (2018). Urbanization in Iron Age Europe: Trajectories, patterns, and social dynamics. *Journal of Archaeological Research*, 26(1), 117-162.

[103] Pearse, P. H. (n.d.). *The Fianna of Fionn*. An Chartlann.

[104] Xu, G. (2018). The Costs of Patronage: Evidence from the British Empire. *American Economic Review*, 108(11), 3170-3198

[105] Encyclopaedia Britannica. (n.d.). Peace of Westphalia. *Encyclopaedia Britannica.*

[106] Sapolsky, Robert M. (2022). *Behave: The Biology of Humans at Our Best and Worst*. Penguin Random House.

[107] Hayes, A. (2024, July 4). Understanding Purchasing Power and the Consumer Price Index. *Investopedia*. Reviewed by M. J. Boyle.

[108] Alden, L. (2023). *Broken Money: Why Our Financial System is Failing Us and How We Can Make it Better*. Lyn Alden.

[109] World Gold Council. (2023). Gold Demand Trends Full Year 2022. *World Gold Council.*

[110] Encyclopaedia Britannica. (n.d.). Gresham's Law. *Encyclopaedia Britannica.*

[111] Snyderman, G. S. (1954). The Functions of Wampum. *Proceedings of the American Philosophical Society*, 98(6), 469-494. University of Pennsylvania Press.

[112] Fitzpatrick, S. M. (2008). Maritime interregional interaction in Micronesia: Deciphering multi-group contacts and exchange systems through time. *Journal of Anthropological Archaeology*, 27(1), 131-147.

[113] Rowlatt, J. (2013, December 8). Why do we value gold? *BBC World Service.*

[114] Encyclopaedia Britannica. (n.d.). Seigniorage. *Encyclopaedia Britannica.*

[115] Williams, J. H. (1941). Deficit Spending. *The American Economic Review*, 30(5), 52-66. American Economic Association.

[116] Britannia Coin Company. (2020, June 29). Counterfeiting, Coin Clipping And The Great Recoinage of 1696. *Britannia Coin Company.*

[117] Helmenstine, A. M. (2019, November 3). What Is a Base Metal? Definition and Examples. *ThoughtCo.*

[118] Encyclopaedia Britannica. (n.d.). Gresham's Law. *Encyclopaedia Britannica.*

[119] Chown, John F. (1994). *A History of Money*. Routledge.

[120] Green, D. B. (2013, November 17). 1278: All Jews of England Arrested in 'Coin-clipping' Scandal. *Haaretz.*

[121] The Holy Bible, English Standard Version. Deuteronomy 23:19.

[122] Green, D. B. (2013, November 17). 1278: All Jews of England Arrested in 'Coin-clipping' Scandal. *Haaretz.*

[123] Baker, J. (2019). The Ecclesiastical Courts. *In Introduction to English Legal History* (5th ed., pp. 135-144). Oxford University Press.

[124] Encyclopaedia Britannica.(n.d.). Windsor Castle. *Encyclopaedia Britannica.*

[125] U.S. Department of the Treasury. (n.d.). "Troubled Assets Relief Program (TARP)." U.S. Department of the Treasury.

[126] Bank of England. (n.d.). Quantitative Easing. *Bank of England.*
[127] Bernanke, Ben S. (2016). What Tools Does the Fed Have Left? Part 3: Helicopter Money. *Brookings Institution.*
[128] Orrell, D. (2019, August 20). A brief history of the international gold standard. *World Finance.*
[129] Merriam-Webster. (n.d.) Fiat. Merriam-Webster.
[130] Bernanke, B. S. (1994). *The Macroeconomics of the Great Depression: A Comparative Approach* (Working Paper No. 4814). National Bureau of Economic Research.
[131] Alden, L. (2023). *Broken Money: Why Our Financial System is Failing Us and How We Can Fix It.* Lyn Alden.
[132] Lamoreaux, N., & Shapiro, I. (Eds.). (2019). *The Bretton Woods Agreements: Together with Scholarly Commentaries and Essential Historical Documents.* Yale University Press.
[133] Calomiris, Charles W., and Stephen H. Haber. (2014). *Fragile by Design: The Political Origins of Banking Crises and Scarce Credit.* Princeton University Press.
[134] Ghizoni, S. K. (1971, August). Nixon Ends Convertibility of U.S. Dollars to Gold and Announces Wage/Price Controls. *Federal Reserve History.* Federal Reserve Bank of Atlanta.
[135] The White House. (n.d.). Richard M. Nixon. *The White House.*
[136] Ghizoni, S. K. (1971, August). Nixon Ends Convertibility of U.S. Dollars to Gold and Announces Wage/Price Controls. *Federal Reserve History.* Federal Reserve Bank of Atlanta.
[137] Meier, G. M. (1977). The Jamaica Agreement, International Monetary Reform, and the Developing Countries. *Journal of International Law and Economics*, 11(1), 67-88.
[138] Bordo, M. D. (2003). *Exchange Rate Regime Choice in Historical Perspective* (Working Paper No. 9654). National Bureau of Economic Research.
[139] Bertaut, Carol, et al. (2021). The International Role of the U.S. Dollar. *FEDS Notes.* Board of Governors of the Federal Reserve System.
[140] Encyclopaedia Britannica. (n.d.). Mutual Assured Destruction. *Encyclopaedia Britannica.*
[141] Tran, H. (2024, June 20). Is the end of the petrodollar near? *Atlantic Council.*
[142] U.S. Department of State.. (n.d.). Oil Embargo, 1973-1974. *Office of the Historian, U.S. Department of State.*
[143] Willner, S. E. (2018). The 1975 Congressional Feasibility Study on "Oil Fields as Military Objectives": U.S.–Saudi Arabian relations and the repercussions of the 1973 Oil Crisis. *The Journal of the Middle East and Africa*, 9(2), 121-136.
[144] Encyclopaedia Britannica. (n.d.). Six-Day War. *Encyclopaedia Britannica.*
[145] Siniver, A. (2008). *Nixon, Kissinger, and U.S. foreign policy making: The machinery of crisis.* Cambridge University Press.

[146] Greenwald, Glenn. (2020, May, 21). The CIA's Murderous Practices, Disinformation Campaigns, and Interference in Other Countries Still Shape the World Order and U.S. Politics. *The Intercept.*

[147] Rotten Tomatoes TV. (2021, June, 18). *"I Am the One Who Knocks!"* Breaking Bad, season 4, episode 6, directed by Michael Slovis, written by Vince Gilligan, aired on AMC, 2011. [Video]. YouTube.

[148] Mankiw, N. G. (2019, December 12). *A skeptic's guide to modern monetary theory.* Prepared for the AEA Meeting, January 2020. Session: Is United States deficit policy playing with fire?

[149] Encyclopaedia Britannica. (n.d.). Hyperinflation in the Weimar Republic. *Encyclopaedia Britannica.*

[150] Federal Reserve. (n.d.). Shares of gross domestic product: Personal consumption expenditures. *Federal Reserve Bank of St. Louis.*

[151] Graeber, David. (2018) *Bullshit Jobs: A Theory.* Simon & Schuster.

[152] Arslanalp, S. and Simpson-Bell, Chima. (2021, May, 5). US Dollar Share of Global Foreign Exchange Reserves Drops to 25-Year Low. *IMF Blog.*

[153] Corporate Finance Institute. (n.d.). Liquidity. *Corporate Finance Institute.*

[154] Board of Governors of the Federal Reserve System. (n.d.). "How does the Federal Reserve affect inflation and employment?" *Board of Governors of the Federal Reserve System.*

[155] Engemann, K. M. (2019, January 16). The Fed's inflation target: Why 2 percent? *Federal Reserve Bank of St. Louis.*

[156] Legal Information Institute. (n.d.). Legal Tender. *Legal Information Institute, Cornell Law School.*

[157] Bachman, Daniel (2024, February 29). A primer on US defense spending. *Deloitte Global Economics Research Center.*

[158] U.S. Department of the Treasury. (1974, September 20). *Inspection of gold at Fort Knox.* [Press release].

[159] Beagley, Alice (2021, February 19). Why Was the Bank of England Founded? *Bank of England Museum.*

[160] U.S. News & World Report. (n.d.). The 25 Best Countries for Economic Stability. *U.S. News & World Report.*

[161] Wright, R. (2020, June 4). Is America becoming a banana republic? *The New Yorker.*

[162] U.S. Government Accountability Office. (2023, May). *The nation's fiscal health: Road map needed to address projected unsustainable debt levels* (GAO-23-106201). Annual Report to Congress.

[163] The Carter Center. (2008, March 3). *President Jimmy Carter - 'Crisis of Confidence' Speech.* [Video]. YouTube.

[164] McKee, M., Karanikolos, M., Belcher, P., & Stuckler, D. (2012). Austerity: A failed experiment on the people of Europe. *Clinical Medicine*, 12(4), 346–350.

[165] Warburton, C. (1944). Monetary expansion and the inflationary gap. *The American Economic Review*, 34(2, Part 1), 303-327.

[166] White, D. (2019). Bankruptcy, morality & student loans: A decade of error in undue hardship analysis. *Ohio Northern University Law Review*, 43(1), 221-256.
[167] Digital History. (n.d.). Why it happened (ID 3432). *University of Houston*.
[168] Cochran, J. P., & Call, S. T. (1998). The role of fractional-reserve banking and financial intermediation in the money supply process: Keynes and the Austrians. *The Quarterly Journal of Austrian Economics*, 1(3), 29–40.
[169] Federal Reserve. (n.d.). Fractional Reserve Banking. *Federal Reserve Bank of Atlanta*.
[170] Federal Reserve. (n.d.). Demand Deposits. *Federal Reserve Bank of St. Louis*.
[171] Federal Reserve. (n.d.). Reserve Requirements. *Federal Reserve*.
[172] Prudential Regulation Authority. (n.d.). Internal Capital Adequacy Assessment Process (ICAAP). *Prudential Regulation Authority*.
[173] Saalmuller, L. (2022, May 19). Cost of capital: What it is & how to calculate. *Harvard Business School Online*.
[174] Shostak, F. (2023, September 13) Why the Fed's Tight Rate Stance Damages the Economy. *Mises Institute*.
[175] Encyclopaedia Britannica. (n.d.). States' Rights. *Encyclopaedia Britannica*.
[176] Encyclopaedia Britannica. (n.d.). Brexit. *Encyclopaedia Britannica*.
[177] Encyclopaedia Britannica. (n.d.). Confederation. *Encyclopaedia Britannica*.
[178] Encyclopaedia Britannica. (n.d.). Federation. *Encyclopaedia Britannica*.
[179] Calomiris, Charles W., and Stephen H. Haber. *Fragile by Design: The Political Origins of Banking Crises and Scarce Credit*. Princeton University Press, 2014.
[180] Colbert, T.B.. (1978). The Populist Moment: A Short History of the Agrarian Revolt in America. *Journal of Economic History*, 39(3), 641-643. Cambridge University Press.
[181] Gallman, R. E., & Wallis, J. J. (Eds.). (1992). *American economic growth and standards of living before the Civil War*. National Bureau of Economic Research (NBER).
[182] Calomiris, Charles W., and Stephen H. Haber. *Fragile by Design: The Political Origins of Banking Crises and Scarce Credit*. Princeton University Press, 2014.
[183] Taylor, Christopher. (2022). *Sterling Volatility and European Monetary Union. Discussion Paper No. 197*. National Institute of Economic and Social Research.
[184] Calomiris, Charles W., and Stephen H. Haber. *Fragile by Design: The Political Origins of Banking Crises and Scarce Credit*. Princeton University Press, 2014.
[185] Vartanian, Thomas P. (2015). *200 Years of American Financial Panics: Crashes, Recessions, Depressions, and the Technology that Will Change It All*. Thomas Vartarian.
[186] Schuler, K. (1992). The world history of free banking: An overview. In *Experience of free banking* (1st ed.). Routledge.

[187] Alden, L. (2023) *Broken Money: Why Our Financial System is Failing Us and How We Can Fix It.* Lyn Alden.
[188] Calomiris, Charles W., and Stephen H. Haber. *Fragile by Design: The Political Origins of Banking Crises and Scarce Credit.* Princeton University Press, 2014.
[189] Encyclopaedia Britannica. (n.d.). Holodomor. *Encyclopaedia Britannica.*
[190] Encyclopaedia Britannica. (n.d.). Cultural Revolution. *Encyclopaedia Britannica.*
[191] Dolar, V. (2022, December 12). The Fed's target for inflation is a made-up number that lacks any concrete evidence. That's kind of the point. *Fortune.*
[192] Christiano, Lawrence J., Martin Eichenbaum, and Charles L. Evans. (1997). *Modeling Money. Working Papers Series, Macroeconomic Issues*, Research Department, Federal Reserve Bank of Chicago. WP-97-17.
[193] Friedman, M. (2006). *The optimum quantity of money.* (1st ed.). Routledge.
[194] World Gold Council. (2023). Gold Demand Trends Full Year 2022. *World Gold Council.*
[195] Jahan, S., Mahmud, A. S., & Papageorgiou, C. (2014). What is Keynesian economics? *Finance & Development*, 51(3).
[196] Encyclopaedia Britannica. (n.d.). Ludwig von Mises. *Encyclopaedia Britannica.*
[197] Keynes, J. M. (1936). *The general theory of employment, interest, and money.* Palgrave MacMillan.
[198] Encyclopaedia Britannica. (n.d.). Quantity Theory of Money. *Encyclopaedia Britannica.*
[199] Stackatoshi Nakamoto. (2024, May 12). *Michael Saylor puts journalist in het [sic] place about Bitcoin and being fiat poor.* [Video]. YouTube.
[200] DeSilver, D. (2018, August 7). For most U.S. workers, real wages have barely budged in decades. *Pew Research Center.*
[201] Cambridge Dictionary. (n.d.). Asset. *Cambridge Dictionary.*
[202] Federal Deposit Insurance Corporation. (2024). Insured institution performance. *FDIC Quarterly*, 18(1).
[203] Adrian, T. (2023, October 10). Higher-for-longer interest rate environment is squeezing more borrowers. *IMF Blog.*
[204] Alden, Lyn. (n.d.). What is Money, Anyway? *LynAlden.com.*
[205] Investopedia. (n.d.). Price-Earnings Ratio (P/E Ratio). *Investopedia.*
[206] Hanlon, S. (2023, November 30). Mediocre 493 may begin to match magnificent 7. *Forbes.*
[207] New World Encyclopedia. (n.d.). Richard Cantillon. *New World Encyclopedia.*
[208] Encyclopaedia Britannica. (n.d.). Quantity Theory of Money. *Encyclopaedia Britannica.*
[209] Encyclopaedia Britannica. (n.d.). Solution (Chemistry). *Encyclopaedia Britannica.*
[210] Encyclopaedia Britannica. (n.d.). Economies of Scale. *Encyclopaedia Britannica.*

[211] Milano, J. (2024, April 19). Why centimillionaires need family offices. *Forbes*.
[212] BlackRock. (2024). Setting the Record Straight: Buying Houses. *BlackRock*.
[213] Encyclopaedia Britannica. (n.d.). NIMBY (Not In My Back Yard). *Encyclopaedia Britannica*.
[214] Bogle, J. C. (2010). *Common sense on mutual funds* (Updated 10th Anniversary ed.). Wiley.
[215] Federal Reserve. (n.d.). Money Stock Measures - M2. *Federal Reserve Bank of St. Louis*.
[216] Measuring Worth. (n.d.). Value of $400 from 2023 to 1995. *Measuring Worth*.
[217] Measuring Worth. (n.d.). Value of $5000 from 2023 to 1995. *Measuring Worth*.
[218] Inequality.org. (n.d.). Wealth Inequality in the United States. *Inequality.org*.
[219] Russell Investments. (2024, February 21). Market concentration and the magnificent seven: Where next? *Seeking Alpha*.
[220] Investopedia. (n.d.). Irrational Exuberance. *Investopedia*.
[221] Encyclopaedia Britannica. (n.d.). Hedging. *Encyclopaedia Britannica*.
[222] LibreTexts. (n.d.). Basic Concepts of Probability. *LibreTexts*.
[223] Encyclopaedia Britannica. (n.d.). Portfolio Asset Allocation: 60-40. *Encyclopaedia Britannica*.
[224] Ghizoni, S. K. (1971, August). Nixon Ends Convertibility of U.S. Dollars to Gold and Announces Wage/Price Controls. *Federal Reserve History*. Federal Reserve Bank of Atlanta.
[225] Chong, Y. D. (2021) Complex Derivatives. *Nanyang Technological University*.
[226] Investopedia. (n.d.). Derivatives 101. *Investopedia*.
[227] Lempérière, Y., Deremble, C., Nguyen, T. T., Seager, P., Potters, M., & Bouchaud, J. P. (2016). Risk premia: asymmetric tail risks and excess returns. *Quantitative Finance*, 17(1), 1–14.
[228] Hammoudeh, S., & McAleer, M. (2013). Risk management and financial derivatives: An overview. *The North American Journal of Economics and Finance*, 25, 109-115.
[229] Greenspan, Alan. (1999, March). Speech at the Boston College Conference on the New Economy, Boston, Massachusetts. *Federal Reserve*.
[230] Stewart, J. B. (2009, September 14). Eight days. *The New Yorker*.
[231] Encyclopaedia Britannica. (n.d.). Stock Market Crash of 1929. *Encyclopaedia Britannica*.
[232] Encyclopaedia Britannica. (n.d.). What Is Greater Fool Theory? *Encyclopaedia Britannica*.
[233] Investopedia. (n.d.). What Is a Bubble? *Investopedia*.
[234] Webb, David Rogers. (2024). *The Great Taking*. David Rogers Webb.
[235] Sastry, P., & Wessel, D. (2015, January 21). The Hutchins Center explains: Quantitative easing. *Brookings Institution*.
[236] Federal Reserve. (n.d.). Money Stock Measures - M1. *Federal Reserve Bank of St. Louis*.
[237] Webb, David Rogers. (2024). *The Great Taking*. David Rogers Webb.

[238] Financial Stability Board. (n.d.). Global Systemically Important Financial Institutions (G-SIFIs). *Financial Stability Board*.
[239] Revill, J. (2023, July 16). Credit Suisse inquiry will keep files secret for 50 years. *Reuters*.
[240] Financial Stability Board. (2023, October 10). 2023 bank failures: Preliminary lessons learnt for resolution. *Financial Stability Board*.
[241] Federal Deposit Insurance Corporation. (n.d.). Chronology of Selected Banking Laws. *Federal Deposit Insurance Corporation*.
[242] Poppers, S. E. (2002, January). *The Austrian theory of business cycles: Old lessons for modern economic policy?* IMF Working Paper No. WP/02/2.
[243] Shostak, F. (2021, September 22). Before a bust, there is always a boom (and malinvestment). *Mises Wire*.
[244] Nye, J. V. C. (n.d.). Standards of living and modern economic growth. *Econlib*.
[245] Stanley, A. (2022, March). Global inequalities. *International Monetary Fund*.
[246] Encyclopaedia Britannica. (n.d.). Peak Oil Theory. *Encyclopaedia Britannica*.
[247] Bernanke, B., Reinhart, V., & Sack, B. (2004). Monetary policy alternatives at the zero bound: An empirical assessment. *Brookings Papers on Economic Activity*, 2004(2), 1-100. Brookings Institution Press.
[248] Daley, J. (2017, January 31). This 100-million-year-old insect trapped in amber defines new order. *Smithsonian Magazine*.
[249] Dowd, K., Hutchinson, M., & Kerr, G. (2012). The coming fiat money cataclysm and the case for gold. *Cato Journal*, 32(2), 363-384.
[250] The Holy Bible. 1 Timothy 6. *New International Version*.
[251] Mark, J.J. (2020). Lao Tzu. *World History Encyclopedia*.
[252] Encyclopaedia Britannica. (2024). Seven deadly sins. *Encyclopaedia Britannica*.
[253] Forbes, J. D. (2011). *Columbus and other cannibals: The wetiko disease of exploitation, imperialism, and terrorism*. Penguin Books.
[254] Merriam-Webster. (n.d.). Greed. *Merriam-Webster*.
[255] Rand, A. (1966). Capitalism: The unknown ideal. Signet Books.
[256] DuBois, C. (1935). *Wintu ethnography*. University of California Publications in American Archaeology and Ethnology.
[257] Investment Master Class. (2023). Charlie Munger: 10 quotes. *Investment Master Class*.
[258] Ballantyne, N., & Ditto, P. H. (2021). Hanlon's razor. *Midwest Studies in Philosophy*, 45(1), 309-331.
[259] Baid, G. (2020). *The joys of compounding*. Columbia University Press.
[260] Merriam-Webster. (n.d.). Greed. *Merriam-Webster*.
[261] Merriam-Webster. (n.d.). Selfish. *Merriam-Webster*.
[262] Merriam-Webster. (n.d.). Excessive. *Merriam-Webster*.
[263] Merriam-Webster. (n.d.). Need. *Merriam-Webster*.
[264] Yang, H. (2021, January 11). *Elon Musk's $195 billion net worth visualized with rice*. [Video]. YouTube.
[265] Tornetta v. Musk, C.A. No. 2018-0408-KSJM (Del. Ch. Jan. 30, 2024) (post-trial opinion).

[266] Legal Information Institute. (n.d.). Dicta. In *Cornell Law School*.
[267] Legal Information Institute. (n.d.). Fact finder. In *Cornell Law School*.
[268] Lakshmi, R. B. (2023, January 11). The environmental impact of battery production for electric vehicles. *Crisis - Biosystem Viability, Global Commons*, Earth.org.
[269] Zehner, O. (2012, October 1). *Green illusions*. [Video]. YouTube. The University of British Columbia.
[270] Emery, D. (2016, December 29). Did Sinclair Lewis say this about fascism in America? *Snopes.com*.
[271] Ncsudreamweaver [user name]. (2019). Ranking the main villains from all 25 official James Bond movies. IMDb.
[272] Young, K. (2023, August 1). A look inside Elon Musk's tiny $50,000 house. *Home & Texture*.
[273] The Infographics Show. (2022, January 20). *The real problem with living on Mars*. [Video]. YouTube.
[274] Heath, F. E. (n.d.). Invisible hand. *Encyclopaedia Britannica*.
[275] Ariely, D. (2010). *Predictably irrational: The hidden forces that shape our decisions* (Revised and expanded ed.). Harper Perennial.
[276] Starlink. (n.d.). *Starlink*.
[277] Tingley, B. (2023, October 2). SpaceX wins $70 million Space Force contract for Starshield military satellites. *Space.com*.
[278] Conger, K., & Hirsch, L. (2022, October 27). Elon Musk completes $44 billion deal to own Twitter. *The New York Times*.
[279] The Boring Company. (n.d.). *The Boring Company*.
[280] Tesla. (n.d.). *Tesla energy*.
[281] Financial Stability Board. (n.d.). *Global systemically important financial institutions (G-SIFIs)*.
[282] Glass, A. (2008, December 19). Bush bails out U.S. automakers. *Politico*.
[283] Kelly, J. (2023, August 18). Follow the incentives and that will tell you everything you need to know about a company's culture. *Forbes*.
[284] Johnson, L. (2024, June 13). Are big businesses really in control? *Score.org*.
[285] Packard, S. (2012, May 5). *Ayn Rand interviewed by Phil Donahue* [Video]. YouTube.
[286] Van Creveld, M. (2010). War in complex environments: The technological dimension. *PRISM*, 1(3).
[287] Encyclopaedia Britannica. (n.d.). Mutual assured destruction. *Encyclopaedia Britannica*.
[288] Encyclopaedia Britannica. (n.d.). Free will. *Encyclopaedia Britannica*.
[289] Glöckner, A. (2023). The irrational hungry judge effect revisited: Simulations reveal that the magnitude of the effect is overestimated. *Judgment and Decision Making,* 11(6). Cambridge University Press.
[290] Behall, K. M., Scholfield, D. J., Hallfrisch, J. G., Kelsay, J. L., & Reiser, S. (1984). Seasonal variation in plasma glucose and hormone levels in adult men and women. *The American Journal of Clinical Nutrition*, 40(6), 1352-1356.

[291] Stanford Iranian Studies Program. (2018, June 12). *Robert Sapolsky: The biology of humans at our best and worst* [Video]. YouTube.
[292] *Id.*
[293] Griffin, G. E. (2002). *The creature from Jekyll Island: A second look at the Federal Reserve* (4th ed.). American Media.
[294] Willis, H. P. (1923). *The Federal Reserve System*. The Ronald Press Company.
[295] Encyclopaedia Britannica. (n.d.). Greenback movement. *Encyclopaedia Britannica*.
[296] Willis, H. P. (1923). *The Federal Reserve System*. The Ronald Press Company.
[297] Collier, I. (2017, October 7). Chicago Ph.D. alumnus and Columbia professor of banking, Henry Parker Willis. *Economics in the rear-view mirror*.
[298] Calomiris, C. W., & Haber, S. (2014). *Fragile by design: The political origins of banking crises and scarce credit*. Princeton University Press.
[299] Encyclopaedia Britannica. (n.d.). The U.S. money market – unit banking system. *Encyclopaedia Britannica*.
[300] Willis, H. P. (1923). *The Federal Reserve System*. The Ronald Press Company.
[301] Calomiris, C. W., & Haber, S. (2014). *Fragile by design: The political origins of banking crises and scarce credit*. Princeton University Press.
[302] National Geographic Society. (n.d.). Hadza. *National Geographic Education*.
[303] The Historical Hierophant. (2023, April 17). *What happened to the last European hunter gatherers?* [Video]. YouTube.
[304] PowerfulJR. (2017, May 16). *Joe Rogan Experience #961 - Graham Hancock, Randall Carlson & Michael Shermer* [Video]. YouTube.
[305] The Historical Hierophant. (2024, April 7). *Magdalenian culture* [Video]. YouTube.
[306] White, D. L. (2016). God, guns & money: A global perspective on intentional homicide. *Western State Law Review*, 44(2).
[307] Gallagher, S. (2019, April 21). What can hunter-gatherers teach us about staying healthy? *Duke Global Health Institute*.
[308] Cox, M. P., Morales, D. A., Woerner, A. E., Sozanski, J., Wall, J. D., & Hammer, M. F. (2009). Autosomal resequence data reveal Late Stone Age signals of population expansion in sub-Saharan African foraging and farming populations. *PLOS One*.
[309] Hammond, P. J. (n.d.). *Rationality in economics*. Department of Economics, Stanford University.
[310] ScienceDirect. (n.d.). Normative theory. *ScienceDirect*.
[311] ScienceDirect. (n.d.). Falsifiability. *ScienceDirect*.
[312] Encyclopaedia Britannica. (n.d.). Karl Popper. *Encyclopaedia Britannica*.
[313] Stanford Encyclopedia of Philosophy. (n.d.). Metaphysics. *Plato.Stanford.edu*.
[314] Côté, D., Caron, A., Aubert, J., et al. (2003). Near wins prolong gambling on a video lottery terminal. *Journal of Gambling Studies*, 19(4), 433-438.
[315] Harrigan, K. A. (2008). Slot machine structural characteristics: Creating near misses using high award symbol ratios. *International Journal of Mental Health and Addiction*, 6(3), 353-368.

[316] Finlay, K., Kanetkar, V., & Marmurek, H. H. C. (2006). The physical and psychological measurement of gambling environments. *Environment and Behavior*, 38(4), 570-581.
[317] Strange, S. (2016). *Casino capitalism* (M. Watson, Intro.). Manchester University Press.
[318] *Id.*
[319] Goodreads. (n.d.). Seneca: Luck is what happens when preparation meets opportunity. *Goodreads*.
[320] Shmaltz and Menudo. (2019, March 17). Famous sayings #141 — 'Luck is when preparation meets …'. *Shmaltz and Menudo*.
[321] BrainyQuote. (n.d.). Samuel Goldwyn: The harder I work, the luckier I get. *BrainyQuote*.
[322] Encyclopaedia Britannica. (n.d.). Seven deadly sins. *Encyclopaedia Britannica*.
[323] Gleick, J. (2008). *Chaos: Making a new science*. Penguin Random House.
[324] ScienceDirect. (n.d.). Causal inference. *ScienceDirect*.
[325] Ridgway, J. D. (2010-2011). Patternicity and persuasion: Evolutionary biology as a bridge between economic and narrative analysis in the law. *Southern Illinois University Law Journal*, 35, 269-315.
[326] Blain, S. D., Longenecker, J. M., Grazioplene, R. G., Klimes-Dougan, B., & DeYoung, C. G. (2020). Apophenia as the disposition to false positives: A unifying framework for openness and psychoticism. *Journal of Abnormal Psychology*, 129(3), 279-292.
[327] ScienceDirect. (n.d.). Causal inference. *ScienceDirect*.
[328] Legal Information Institute. (n.d.). But-for test. In *Cornell Law School*.
[329] Carroll, L. (1865). *Alice's adventures in wonderland* (Project Gutenberg). Carnegie Mellon University.
[330] Encyclopaedia Britannica. (n.d.). John Wilkes Booth. *Encyclopaedia Britannica*.
[331] *Palsgraf v. Long Island Railroad Co.*, 248 N.Y. 339 (N.Y. 1928).
[332] Merriam-Webster. (n.d.). Heuristic. *Merriam-Webster*.
[333] Vernon, J. L. (n.d.). Understanding the butterfly effect. *American Scientist*.
[334] IMDb. (n.d.). Havana (2004). *IMDb*.
[335] IMDb. (n.d.). The Butterfly Effect (2004). *IMDb*.
[336] Hirshleifer, J. (2001). *The dark side of the force: Economic foundations of conflict theory*. Cambridge University Press.
[337] Nooah. (2024, May 27). *Jeff Bezos - About luck* [Video]. YouTube.
[338] Hendrix, L. (2021, November 18). *Millionaire goes homeless to prove anyone can make $1,000,000* [Video]. YouTube.
[339] Kang, W., Guzman, K. L., & Malvaso, A. (2023). Big Five personality traits in the workplace: Investigating personality differences between employees, supervisors, managers, and entrepreneurs. *Frontiers in Psychology*, 14, 976022.
[340] Eldridge, S. (n.d.). Survivorship bias. *Encyclopaedia Britannica*.
[341] TJ3 History. (2021, October 29). *How a mathematician saved the American bombers in World War II - Abraham Wald & survivorship bias* [Video]. YouTube.

[342] Wiegmann, D. A., Wood, L. J., Cohen, T. N., & Shappell, S. A. (2022). Understanding the "Swiss Cheese Model" and its application to patient safety. *Journal of Patient Safety*, 18(2), 119-123.
[343] PRIDE MMA. (2020, May 28). *Conor vs Aldo - Full fight* [Video]. YouTube.
[344] UsingEnglish.com. (n.d.). Cream rises to the top - idiom. *UsingEnglish.com*.
[345] Frank, R. H. (2016). *Success and luck: Good fortune and the myth of meritocracy*. Princeton University Press.
[346] Encyclopaedia Britannica. (n.d.). Normal distribution. *Encyclopaedia Britannica*.
[347] Frank, R. H. (2016). *Success and luck: Good fortune and the myth of meritocracy*. Princeton University Press.
[348] Encyclopaedia Britannica. (n.d.). Hindsight bias. *Encyclopaedia Britannica*.
[349] Blackstone, W. (1765). *Commentaries on the laws of England: Book the first*. Clarendon Press.
[350] IMDb. (n.d.). *Downton Abbey* (2010-2015). IMDb.
[351] Encyclopaedia Britannica. (n.d.). Highclere castle. *Encyclopaedia Britannica*.
[352] Highclere Castle. (n.d.). The estate. *Highclere Castle*.
[353] Corbin, T. (2021, January 23). Downtown Abbey: 30 facts every fan needs to know about Highclere Castle. *Hampshire Live*.
[354] Highclere Castle. (n.d.). The estate. *Highclere Castle*.
[355] Nearly Noble. (n.d.). All UK Country Houses and Stately Homes. *Nearly Noble*.
[356] Goldbart, M. (2022, September 20). Queen Elizabeth II's funeral watched by 37.5M viewers in UK. *Deadline*.
[357] The Crown Estate. (n.d.). Annual report. *The Crown Estate*.
[358] Serhan, Y. (2023, April 28). Here's what we know about the vast cost of King Charles III's coronation. *Time*.
[359] Blackstone, W. (1765). *Commentaries on the laws of England: Book the first*. Clarendon Press.
[360] Library of Congress. (n.d.). With malice towards none: The Abraham Lincoln Bicentennial Exhibition. Practicing law. *Library of Congress*.
[361] Steiner, M.E. (2009). Lincoln's law books. Marquette Law School.
[362] Lancaster University. (n.d.). *Conceptualizations of animals, plants and nature in the West: an historical approach*. Lancaster University.
[363] *Id*.
[364] gregorija1. (2008, July 10). *Star Trek racism* [Video]. YouTube.
[365] Tempelman, J. H. (2010). Austrian business cycle theory and the global financial crisis: Confessions of a mainstream economist. *The Quarterly Journal of Austrian Economics*, 13(1), 3-15.
[366] Weinberg, J. (n.d.). *The Great Recession and its aftermath (2007–)*. Federal Reserve Bank of Richmond.
[367] National Commission on the Causes of the Financial and Economic Crisis in the United States. (2011). *Final report of the National Commission on the Causes of the Financial and Economic Crisis in the United States*. U.S. Government Printing Office.

368 Rothwell, J. (2020, September 8). *Assessing the economic gains of eradicating illiteracy nationally and regionally in the United States.* Barbara Bush Foundation, Gallup.

369 Taibbi, M. (2010, April 5). The great American bubble machine. *Rolling Stone.*

370 Taibbi, M. (2009, April 2). How Wall Street is using the bailout to stage a revolution. *Rolling Stone.*

371 Federal Reserve History. (n.d.). Glass-Steagall Act. *Federal Reserve History.*

372 Emmons, B. (2009). *Reader exchange: Warren Buffett: Financial Weapons of Mass Destruction.* Federal Reserve Bank of St. Louis.

373 Webb, D. R. (2023). *The great taking* (M. Palmer, Ed.). David Rogers Webb.

374 Taibbi, M. (2009, April 2). How Wall Street is using the bailout to stage a revolution. *Rolling Stone.*

375 Concodanomics. [@concodanomics]. (n.d.). *Profile.* X.

376 Nooah. (2024, May 27). *Jeff Bezos - About luck* [Video]. YouTube.

377 Feeding America. (n.d.). Food waste statistics in the US. *Feeding America.*

378 Rivera, J. L., & Lallmahomed, A. (2015). Environmental implications of planned obsolescence and product lifetime: A literature review. *International Journal of Sustainable Engineering*, 9(2), 119-129.

379 Turrentine, J. (2019, July 12). The United States is the most wasteful country in the world: We're number one! We're number one! *NRDC.*

380 United Nations. (n.d.). Ending poverty. *United Nations.*

381 World Health Organization. (n.d.). Child mortality and causes of death. *The Global Health Observatory.*

382 Talbot, D. (2012, October 29). Given tablets but no teachers, Ethiopian children teach themselves. *MIT Technology Review.*

383 White, D. L. (2016). The Anti-Money Laundering Complex in the Modern Era. *The Banking Law Journal*, 133(10).

384 *Id.*

385 Kratter, M. R. (2021). A beginner's guide to Bitcoin. *Matthew R. Kratter.*

386 Meiklejohn, S., Pomarole, M., Jordan, G., Levchenko, K., McCoy, D., Voelker, G. M., & Savage, S. (2016). A fistful of Bitcoins: Characterizing payments among men with no names. *Communications of the ACM*, 59(4), 86-93.

387 Georgiadis, E. (2019). *How many transactions per second can Bitcoin really handle? Theoretically.* (Paper No. 2019/416).

388 Sguanci, C., Spatafora, R., & Vergani, A. M. (2021). Layer 2 blockchain scaling: A survey. *arXiv*:2107.10881. Cornell University.

389 Internal Revenue Service. (n.d.). Digital assets. U.S. Department of the Treasury.

390 Chainalysis. (n.d.). *Chainalysis.*

391 Chaboud, A., Rime, D., & Sushko, V. (2023). *The foreign exchange market* (BIS Working Paper No. 1094). Monetary and Economic Department.

[392]Cones, J. W. (1997). *The feature film distribution deal: A critical analysis of the single most important film industry agreement*. Southern Illinois University Press.

[393]*Id.*

[394]White, D. L. (2016). The Anti-Money Laundering Complex in the Modern Era. *The Banking Law Journal*, 133(10).

[395]International Consortium of Investigative Journalists. (2016). The Panama Papers: Exposing the rogue offshore finance industry. *International Consortium of Investigative Journalists*.

[396]Binance Academy. (n.d.). Transactions per second (TPS). *Binance Academy*.

[397]Nakamoto, S. (2009). Bitcoin: A peer-to-peer electronic cash system. *Bitcoin.org*.

[398]Mill Hill Books. (2019). Kicking the hornet's nest: The complete writings, emails, and forum posts of Satoshi Nakamoto, the founder of Bitcoin and cryptocurrency. *Mill Hill*.

[399]satoshi. (2010, August 27). Re: Bitcoin does NOT violate Mises' regression theorem [Forum post]. BitcoinTalk.

[400]Morris, C. (2023, July 27). Understanding gold's intrinsic value. *ByteTree*.

[401]Voskuil, E. (2020). *Cryptoeconomics: Fundamental principles of Bitcoin* (J. Chiang, Ed. & Illus.). Eric Voskuil.

[402]CoinGecko. (n.d.). Cryptocurrency Prices by Market Cap. *CoinGecko*.

[403]Nakamoto, S. (2009). Bitcoin: A peer-to-peer electronic cash system. *Bitcoin.org*.

[404]Bitcoin.org. (n.d.). Development. *Bitcoin.org*

[405]Pritzker, Y. (2019). *Inventing Bitcoin: The technology behind the first truly scarce and decentralized money explained* (N. Evans, Illus.). Yan Pritzker.

[406]Lieberman, M. B., & Montgomery, D. B. (1988). First-mover advantages. *Strategic Management Journal*, 9 (Special Issue: Strategy Content Research), 41-58.

[407]Alden, L. (2021, March). Analyzing Bitcoin's network effect. *LynAlden.com*.

[408]De Filippi, P., & Loveluck, B. (2016). The invisible politics of Bitcoin: Governance crisis of a decentralized infrastructure. *Internet Policy Review*, 5(4).

[409]Ranganthan, V. P., Dantu, R., Paul, A., Mears, P., & Morozov, K. (2018). A decentralized marketplace application on the Ethereum blockchain. In *2018 IEEE 4th International Conference on Collaboration and Internet Computing* (CIC) (pp. 90-97). Philadelphia, PA: IEEE.

[410]Hanyecz, L. (2010, May 22). First commercial Bitcoin transaction. *Guinness World Records*.

[411]Ethereum Foundation. (n.d.). Welcome to Ethereum. *Ethereum.org*

[412]Messari. (n.d.). What is Ethereum?. *Messari.io*

[413]CoinCodex. (n.d.). Ethereum (ETH) ICO: ICO dates July 22, 2014 - September 2, 2014. *CoinCodex*.

[414]Justia Trademarks. (n.d.). Ether (Serial No. 87791341). *Justia*.

[415]Dowlat, S. (2018, July 11). Cryptoasset market coverage initiation: Network creation. *Satis Research. Bloomberg.*

[416]Duignan, B. (2024, July 29). Dot-com bubble: Stock market [1995–2000]. *Encyclopaedia Britannica.*

[417]Solana. (n.d.). Solana. *solana.com*

[418]Token Terminal. (n.d.). Solana - key metrics. *tokenterminal.com*

[419]Nintendo. (n.d.). About Nintendo. *Nintendo.com*

[420]CompaniesMarketCap. (n.d.). Nintendo revenue. *CompaniesMarketCap.com*

[421]Dowlat, S. (2018, July 11). Cryptoasset market coverage initiation: Network creation. *Satis Research. Bloomberg.*

[422]White, D. (2023, August 6). The quiet part out loud: What crypto influencers never tell you. *Coinmonks.*

[423]U.S. Securities and Exchange Commission. (n.d.). Crypto assets and cyber enforcement actions. *SEC.*

[424]PwC. (2024, June 25). Trends in SEC enforcement actions. US In Depth, 2024-03. *PwC.*

[425]*U.S. Securities and Exchange Commission v. Payward, Inc., et al.*, No. 23-cv-06003-WHO (N.D. Cal. filed 2023).

[426]Mendelson, M. (2019). From initial coin offerings to security tokens: A U.S. federal securities law analysis. *Stanford Technology Law Review*, 22, 52.

[427]Commodity Futures Trading Commission. (n.d.). Bitcoin basics. *CFTC.*

[428]Internal Revenue Service. (n.d.). Digital assets. *U.S. Department of the Treasury.*

[429]Encyclopaedia Britannica. (n.d.). Laissez-faire. *Encyclopaedia Britannica.*

[430]White, D. L. (2015). Title III of the JOBS Act: Congress invites investor abuse and leaves the SEC holding the bag. *Willamette Law Review*, 52(2).

[431]Li, W., Bao, L., Chen, J., Grundy, J., Xia, X., & Yang, X. (2024). Market manipulation of cryptocurrencies: Evidence from social media and transaction data. *ACM Transactions on Internet Technology*, 24(2), Article 8, 1-26.

[432]Mirtaheri, M., Abu-El-Haija, S., Morstatter, F., Van Der Steeg, G., & Galstyan, A. (2021). Identifying and analyzing cryptocurrency manipulations in social media. *IEEE Transactions on Computational Social Systems*, 8(3), 607-617.

[433]Agarwal, S., Atondo-Siu, G., Ordekian, M., Hutchings, A., Mariconti, E., & Vasek, M. (2024). Short paper: DeFi deception—Uncovering the prevalence of rugpulls in cryptocurrency projects. In F. Baldimtsi & C. Cachin (Eds.), *Financial cryptography and data security*. FC 2023 (Lecture Notes in Computer Science, Vol. 13950). Springer, Cham.

[434]Electronic Code of Federal Regulations. (n.d.). *17 CFR Part 227 - Regulation crowdfunding.*

[435]CNC Intelligence. (2024, April 14). Crypto ICO scams: A comprehensive guide. *CNC Intelligence.*

[436]White, D. L. (2015). Title III of the JOBS Act: Congress invites investor abuse and leaves the SEC holding the bag. *Willamette Law Review*, 52(2).

437 Taplin, J. (2017). *Move fast and break things: How Facebook, Google, and Amazon cornered culture and undermined democracy*. Little, Brown & Company.
438 Briola, A., Vidal-Tomás, D., Wang, Y., & Aste, T. (2023). Anatomy of a stablecoin's failure: The Terra-Luna case. *Finance Research Letters*, 51, 103358.
439 International Monetary Fund. (2020, October 19). *Cross-border payment—A vision for the future* [Video]. YouTube.
440 Encyclopaedia Britannica. (n.d.). Big Brother (fictional character). *Encyclopaedia Britannica*.
441 Dinh, V. D. (2002). Foreward - Freedom and security after September 11. *Harvard Journal of Law & Public Policy*, 25, 399.
442 Wikipedia contributors. (n.d.). Supreme leader. *Wikipedia*.
443 Merriam-Webster. (n.d.). Assay. *Merriam-Webster*.
444 Chowdhury, A. (n.d.). The Cantillion effect. *Adam Smith Institute*.
445 Hughes, T. A., & Royde-Smith, J. G. (n.d.). World War II. *Encyclopaedia Britannica*.
446 White, D. L. (2016). The Anti-Money Laundering Complex in the Modern Era. *The Banking Law Journal*, 133(10).
447 *Id.*
448 *Id.*
449 Financial Action Task Force. (n.d.). FATF recommendations. *FATF*.
450 Gouvin, E. J. (2003). Bringing out the big guns: The USA PATRIOT Act, money laundering, and the war on terrorism. *Baylor Law Review*, 55, 955-985.
451 Financial Crimes Enforcement Network. (n.d.). USA PATRIOT Act. *FinCEN*.
452 White, D. L. (2016). The Anti-Money Laundering Complex in the Modern Era. *The Banking Law Journal*, 133(10).
453 Diamond, D. W., & Dybvig, P. H. (1983). Bank runs, deposit insurance, and liquidity. *Journal of Political Economy*, 91(3), 401-419.
454 Richardson, G. (n.d.). Banking panics of 1930-31. *Federal Reserve History*.
455 Fung, K. (2022, February 18). Banks have begun freezing accounts linked to trucker protest. *Newsweek*.
456 Levush, R. (2013, April 4). The Cyprus banking crisis and its aftermath: Bank depositors be aware. *Library of Congress Blogs*.
457 Kiguel, M. (n.d.). Argentina's 2001 economic and financial crisis: Lessons for Europe. In *Think Tank 20: Beyond macroeconomic policy coordination discussions in the G-20*.
458 Bayeh, B., Cubides, E., & O'Brien, S. (2024). Findings from the diary of consumer payment choice. *Federal Reserve Financial Services*.
459 CBS San Francisco. (2013, April 15). Silicon Valley prostitutes using mobile apps to accept credit card payments. *CBS San Francisco*.
460 Nikos-Rose, K. (2023, November 21). Digital payment platforms can easily be misused for drug dealing. *UC Davis News*.
461 World Economic Forum. (2024, April). Financial literacy and money education. *WEF*.

[462] Pew Research Center. (2018, April 26). The public, the political system and American democracy: Political engagement, knowledge and the midterms. *Pew Research Center*.

[463] Rothwell, J. (2020, September 8). Assessing the economic gains of eradicating illiteracy nationally and regionally in the United States. *Barbara Bush Foundation* and *Gallup*.

[464] National Commission on Terrorist Attacks Upon the United States. (2004). *The 9/11 Commission report: Final report of the National Commission on Terrorist Attacks Upon the United States*. U.S. Government Printing Office.

[465] Chen, J. (2024, August 6). Know your client (KYC): What it means and compliance requirements. *Investopedia*.

[466] Frax Finance. (n.d.). AMO overview. *Frax.finance*.

[467] U.S. Department of the Treasury. (n.d.). Troubled Asset Relief Program (TARP). *U.S. Department of the Treasury*.

[468] U.S. Energy Information Administration. (2024, February 1). Tracking electricity consumption from U.S. cryptocurrency mining operations. *U.S. Energy Information Administration*.

[469] Ekins, E., & Gygi, J. (2023, May 31). Poll: Only 16% of Americans support the government issuing a central bank digital currency. *Cato Institute*.

[470] Nakamoto, S. (2009). Bitcoin: A peer-to-peer electronic cash system. *Bitcoin.org*.

[471] U.S. Securities and Exchange Commission. (n.d.). Exchange-traded fund (ETF). *Investor.gov*.

[472] U.S. Securities and Exchange Commission. (2023). *Counterfeiting stock 2.0 - Comments to the SEC*. [PDF].

[473] Frederick, S., Loewenstein, G., & O'Donoghue, T. (2002). Time discounting and time preference: A critical review. *Journal of Economic Literature*, 40(2), 351-401.

[474] Practical Law. (n.d.). Glossary - Trust. *Thomson Reuters*.

[475] Shibata, K. (1931). The subjective theory of value and theories of the value of money. *Kyoto University Economic Review*, 6(1), 71-93.

[476] Maraia, M. (2022, August 26). What does it mean to orange-pill someone? *Bitcoin Magazine*.

[477] Saylor, M. [@saylor]. (n.d.). *Profile*. X.

[478] MicroStrategy. (2020, December 21). *MicroStrategy announces over $1B in total Bitcoin purchases in 2020*. [Press release].

[479] Wikipedia contributors. (n.d.). Flesch–Kincaid readability tests. *Wikipedia*.

[480] Voskuil, E. [@evoskuil]. (n.d.). *Profile*. X.

[481] Bitcoin Takeover. (2024, May 4). *S15 E22: Eric Voskuil on Libbitcoin & Bitcoin culture* [Video]. YouTube.

[482] Voskuil, E. (2020). *Cryptoeconomics: Fundamental principles of Bitcoin* (J. Chiang, Ed. & Illus.). Eric Voskuil.

[483] Financial Stability Board. (n.d.). Global systemically important financial institutions (G-SIFIs). *FSB*.

[484] Statista. (n.d.). Number of renter-occupied housing units in the United States from 1975 to 2023 (in millions). *Statista.*
[485] Robinson, K. J. (n.d.). Savings and loan crisis (1980–1989). *Federal Reserve History.*
[486] World Health Organization. (2020, August 5). Palliative care. *WHO.*
[487] Boaz, D. (n.d.). Libertarianism. *Encyclopaedia Britannica.*
[488] Costa, D. (n.d.). Anarcho-capitalism. *Encyclopaedia Britannica.*
[489] Hughes, E. (n.d.). *A cypherpunk's manifesto.* Activism.net.
[490] National Institute of Allergy and Infectious Diseases. (n.d.). Antiretroviral drug development. *National Institute of Allergy and Infectious Diseases.*
[491] Laslo. (2010, May 18). Bitcoin for pizzas [Forum post]. *Bitcoin Forum on Simple Machines Forum.*
[492] Seyr, R. (2024, March 9). *Bitcoin supercharges your life | Eric V Stacks* [Video]. YouTube.
[493] Booth, J. (n.d.). *[Jeff Booth's website].*
[494] History.com Editors. (2022, October 4). Maginot Line. *History.com.*
[495] Encyclopaedia Britannica. (n.d.). Mutual assured destruction. *Encyclopaedia Britannica.*
[496] Wilson, C. (2020). Artificial intelligence and warfare. In M. Martellini & R. Trapp (Eds.), *21st century Prometheus.* Springer.
[497] TRT World. (2024, January 22). *Connor Leahy - The InnerView* [Video]. YouTube.
[498] Center for Humane Technology. (2023, March 9). *The A.I. dilemma* [Video]. YouTube.
[499] *Murthy, Surgeon General, et al. v. Missouri, et al.*, No. 23-411, Supreme Court of the United States. Argued March 18, 2024—Decided June 26, 2024.
[500] Wikipedia contributors. (n.d.). Ground truth. *Wikipedia.*
[501] Hernandez, J. (2023, March 22). That panicky call from a relative? It could be a thief using a voice clone, FTC warns. *NPR.*
[502] Buchanan, J. M. (2003). The free rider problem. *Stanford Encyclopedia of Philosophy.*

Made in the USA
Columbia, SC
12 October 2024